Functional Materials for Next-Generation Rechargeable Batteries

Functional Materials for Next-Generation Rechargeable Batteries

editors

Jiangfeng Ni
Soochow University, China

Li Lu
National University of Singapore, Singapore

World Scientific

NEW JERSEY · LONDON · SINGAPORE · BEIJING · SHANGHAI · HONG KONG · TAIPEI · CHENNAI · TOKYO

Published by

World Scientific Publishing Co. Pte. Ltd.

5 Toh Tuck Link, Singapore 596224

USA office: 27 Warren Street, Suite 401-402, Hackensack, NJ 07601

UK office: 57 Shelton Street, Covent Garden, London WC2H 9HE

Library of Congress Control Number: 2020950204

British Library Cataloguing-in-Publication Data
A catalogue record for this book is available from the British Library.

FUNCTIONAL MATERIALS FOR NEXT-GENERATION RECHARGEABLE BATTERIES

ISBN 978-981-123-066-0 (hardcover)
ISBN 978-981-123-067-7 (ebook for institutions)
ISBN 978-981-123-068-4 (ebook for individuals)

For any available supplementary material, please visit
https://www.worldscientific.com/worldscibooks/10.1142/12110#t=suppl

Typeset by Stallion Press
Email: enquiries@stallionpress.com

Preface

Over-consumption of fossil fuels has caused deficiency of limited resources and environmental pollution. Hence, deployment and utilization of renewable energy become an urgent need, which has triggered the development of more advanced high-energy and low-cost energy storage systems. Rechargeable batteries beyond lithium-ion such as lithium-sulfur, sodium-ion, and potassium-ion, therefore become feasible solutions. Intensive research works have been devoted to the development of innovative energy storage chemistries, novel concepts of energy storage devices, and new architectures of electrode materials. To address the most state-of-the-art progress, this book is presented. It contains a number of review chapters as well as reports on the recent research results.

This book starts with the principles and fundamentals of lithium rechargeable Batteries (Chapter 1), followed by the advancement of functional materials for next-generation rechargeable batteries which might meet the requirement of high energy density and high safety. Among them, rechargeable lithium-sulfur (Li-S) batteries have received considerable interests, owing to their low-cost and high energy density (2600 Wh kg^{-1}) enabled by multiple electron transfer. In Chapter 2, Zhu and Li summarize the recent progresses in the carbon-metal oxide composites for Li-S battery cathodes, and offer some insights on the future directions of carbon-metal oxide hybrid cathodes for high performance Li-S batteries. Transition-metal sulfides are also interesting cathode materials for Li-S batteries. The challenges and perspectives of transition-metal sulfides have been addressed by Chen *et al.* (Chapter 3). A unique structure of polypyrrole and graphene oxide shows a high specific capacity of 548.4 mAh g^{-1} at a high charge rate of 5.0 C (Chapter 4), while the carbon nanoflakes/sulfur arrays exhibit high capacities of 1117 mAh g^{-1} at 0.2 C, and 741 mAh g^{-1} at 0.6 C with good high-rate cycling performance (Chapter 5).

Recently sodium-ion batteries have been emerging as a rising battery-systems alternative to lithium-ion, but they are facing significant materials and technical challenges. One of the problems is to find a suitable and practically applicable anode material. Two mini-reviews summarize the state-of-the-art progress of hard carbon (Chapter 6) and MoS$_2$ anodes (Chapter 7) for efficient sodium storage, focusing on their electrochemical mechanisms, intrinsic advantages and modulation strategies to improve their performance. Researchers also report on carbon nanoflake (Chapter 8), biomass carbon from tree leaves (Chapter 9), and Fe$_2$O$_3$ nanorods (Chapter 10) showing interesting sodium-storage properties.

Besides sodium-ion, potassium-ion batteries also possess appealing characters of low cost, high ionic conductivity, high power ability, and environmental friendliness. Polythiophene, an organic anode material, shows a reversible capacity of 58 mAh g^{-1} at 30 mA g^{-1}, suggesting that it might be a promising anode material for potassium storage (Chapter 11). Researchers also demonstrate nitrogen-doped MnO$_2$ nanorods for high-energy Zn-MnO$_2$ batteries (Chapter 12). Metal-organic frameworks, consisting of metal ions or clusters coordinated to organic ligands to form one-, two-, or three-dimensional structures, have attracted great interest in energy storage because of their unique porous structures and organic–inorganic hybrid nature. A contribution reviewing metal-organic frameworks derived structures for next-generation rechargeable batteries has been included in Chapter 13.

Besides electrode materials, electrolytes are also a key component of rechargeable batteries, serving as the medium for the transfer of charges. In view of this, a chapter dealing with polymer-in-salt solid electrolytes has also been enclosed in this book (Chapter 14).

We sincerely hope that these aforementioned contributions will help and assist your research in materials and devices on next-generation rechargeable batteries.

Jiangfeng Ni
Soochow University, China

Li Lu
National University of Singapore

Contents

Preface   v

Chapter 1   Principles and Fundamentals of Lithium
              Rechargeable Batteries   1
              Jiangfeng Ni

Chapter 2   Carbon-Metal Oxide Nanocomposites as
              Lithium-Sulfur Battery Cathodes   29
              Sheng Zhu and Yan Li

Chapter 3   Recent Advances of Polar Transition-Metal
              Sulfides Host Materials for Advanced
              Lithium–Sulfur Batteries   47
              Liping Chen, Xifei Li, and Yunhua Xu

Chapter 4   Graphene Oxide-Polypyrrole Composite
              as Sulfur Hosts for High-Performance
              Lithium–Sulfur Batteries   65
              Qian Wang, Chengkai Yang, Hui Tang,
              Kai Wu, and Henghui Zhou

Chapter 5   Synthesis of Carbon Nanoflake/Sulfur
              Arrays as Cathode Materials of
              Lithium–Sulfur Batteries   77
              Fan Wang, Xinqi Liang, Minghua Chen,
              and Xinhui Xia

Chapter 6   Hard Carbon Anode Materials for
               Sodium-Ion Batteries   87
               Ismaila El Moctar, Qiao Ni, Ying Bai,
               Feng Wu, and Chuan Wu

Chapter 7   Some MoS_2-Based Materials for
               Sodium-Ion Battery   111
               Qing Li, Xiaotian Guo, Mingbo Zheng,
               and Huan Pang

Chapter 8   Carbon Nanoflakes as a Promising Anode
               for Sodium-Ion Batteries   127
               Xiaocui Zhu, S. V. Savilov, Jiangfeng Ni,
               and Liang Li

Chapter 9   Phoenix Tree Leaves–Derived Biomass
               Carbons for Sodium-Ion Batteries   135
               Zengqiang Tian, Shijiao Sun, Xiangyu Zhao,
               Meng Yang, and Chaohe Xu

Chapter 10   Flexible α-Fe_2O_3 Nanorod Electrode
               Materials for Sodium-Ion Batteries with
               Excellent Cycle Performance   147
               Depeng Zhao, Di Xie, Hengqi Liu, Fang Hu,
               and Xiang Wu

Chapter 11   Green and Facile Synthesis of Nanosized
               Polythiophene as an Organic Anode for
               High-Performance Potassium-Ion Battery   159
               Guifang Zeng, Yongling An, Huifang Fei,
               Tian Yuan, Sun Qing, Lijie Ci, Shenglin Xiong,
               and Jinkui Feng

Chapter 12   Nitrogen-Doped MnO_2 Nanorods as
               Cathodes for High-Energy Zn-MnO_2 Batteries   167
               Yalan Huang, Wanyi He, Peng Zhang, and
               Xihong Lu

Chapter 13   Metal-Organic Framework–Derived Structures
for Next-Generation Rechargeable Batteries   179

*Wenhui Shi, Xilian Xu, Lin Zhang,
Wenxian Liu, and Xiehong Cao*

Chapter 14   Polymer-in-Salt Solid Electrolytes for
Lithium-Ion Batteries   201

*Chengjun Yi, Wenyi Liu, Linpo Li,
Haoyang Dong, and Jinping Liu*

Chapter 1
Principles and Fundamentals of Lithium Rechargeable Batteries

Jiangfeng Ni[*]

Lithium rechargeable batteries have witnessed significant technical advances in the electrochemical performance and stability, together with success in penetrating into every corner of our lives. This success also leads to the 2019 Nobel Prize in Chemistry being awarded to three scientists for the development of lithium-ion batteries. In this chapter, the principles and fundamentals of lithium rechargeable batteries are reviewed, aiming to provide a beginner's guide to researchers going into battery communities. The basic concepts and characteristics of batteries are expounded, showing how lithium rechargeable batteries are developed and assessed. A brief summary of battery material is then provided, highlighting some key cathode and anode components, whose discovery and optimization finally leads to the success of commercial batteries. Topics discussed also include the battery design and manufacture and offer insights into the future development of lithium rechargeable batteries. The knowledge deposited in this chapter is expected to stimulate the design of functional materials in multiple dimensions and scales required for future battery applications.

---

[*]School of Physical Science and Technology, Center for Energy Conversion Materials & Physics, Soochow University, Suzhou 215006, P. R. China. Email: jeffni@suda.edu.cn

Keywords: Lithium rechargeable batteries; working principle; cathode; anode; electrolyte; separator.

1. Introduction

A battery is a device that converts chemical energy into electrical energy by electrochemical oxidation and reduction reactions occurring at the two electrodes.[1,2] Rechargeable batteries are electric devices capable of charging and discharging. Due to their ability to store large quantities of electricity in a compact form and to discharge current at high output, they are widely used in personal computers, cell phones, tablets, and other mobile devices, and more recently in hybrid electric vehicles (HEVs), electric vehicles (EVs) and stationary power storage systems.[3–5]

Rechargeable batteries are composed of positive (cathode) and negative (anode) electrodes separated by a separator which is filled with an electrolytic solution. For lithium rechargeable batteries (LRBs), the positive electrode is often a lithium-containing compound, and the negative electrode a carbon material. A binder is used to attach them to a current collector (copper foil and aluminum foil). The positive and negative electrode materials and the separator are rolled inside a metal case together with the electrolytic solution. When the battery is connected to an external load or device to be powered, the negative electrode supplies a current of electrons that flow through the load and are accepted by the positive electrode; meanwhile, Li^+ ions are extracted from negative host matrix and inserted into positive electrode materials via the internal electrolyte solution. Since the operation of LRBs involves Li^+ ions extraction/insertion from/into a host matrix, they are often referred to as "rocking chair batteries".

Compared with other rechargeable batteries, LRBs show many advantages that make them extremely attractive in personal electronics. These advantages include (1) High energy densities, ~250 Wh kg^{-1} or ~500 Whl^{-1}; (2) High operating voltages, ~3.6 V and 3-fold that of Ni-Cd and Ni-MH batteries; (3) Excellent rechargeability up to 1,000 to 3,000 cycles; (4) Minimal self-discharge, under 10% per month; (5) Almost no memory effect; (6) Operation over a wide range of temperatures; (7) Last but not least, environmental friendliness. As a result, LRBs have now been widely utilized in portable electronics and small energy storage. However, their scalable applications in large-format systems such as automobile and energy storage are still hindered due to economic and safety concerns.

Lithium battery includes lithium primary battery and LRB, while the latter is also known specifically as Li-ion battery. Experimentation with lithium primary batteries began as early as in 1912 under G.N. Lewis. Many lithium primary batteries such as $Li-MnO_2$, $Li-SO_2$, and $Li-SOCl_2$ were further pursued in the 1950s.[6] From the 1970s these batteries showed application perspective in military fields because of their high work potential and large energy density. At the same time, rechargeable lithium metal battery was developed, and successfully invented and put into the market under the consideration of environmental protection and resources. However, its future development has long been at standstill because of the low cycle efficiency and terrible safety issue.

The concept of LRB was started in the 1970s when Whittingham with Exxon Company demonstrated the chemical intercalation of lithium in the Li_xTiS_2 material.[7] In the 1980s, Goodenough at Oxford University discovered that Li_xCoO_2 could serve as a cathode material, in which the van der Waals gaps between the cobalt dioxide layers allow Li^+ ions for reversible (de)intercalation. The following discovery of the petroleum coke anode enabled Akira Yoshino at Asahi Kasei Corporation to create the first commercially viable LRB in 1985. This LRB was lightweight, hardwearing and could be recharged hundreds of times. The key to this success is that this LRB is based on topotactic reactions rather than chemical reactions, involving only Li^+ ions shuttling between the anode and cathode. The configurations of the reliable prototypes of LRBs are illustrated in Fig. 1. For their pioneering work on the development of lithium-ion batteries, Goodenough, Whittingham, and Yoshino have been awarded the Nobel Prize in Chemistry 2019 by the Royal Swedish Academy of Sciences.[8]

Inspired by their work, Sony initiated the commercialization of LRBs in 1991, which has been of the greatest benefit to humankind

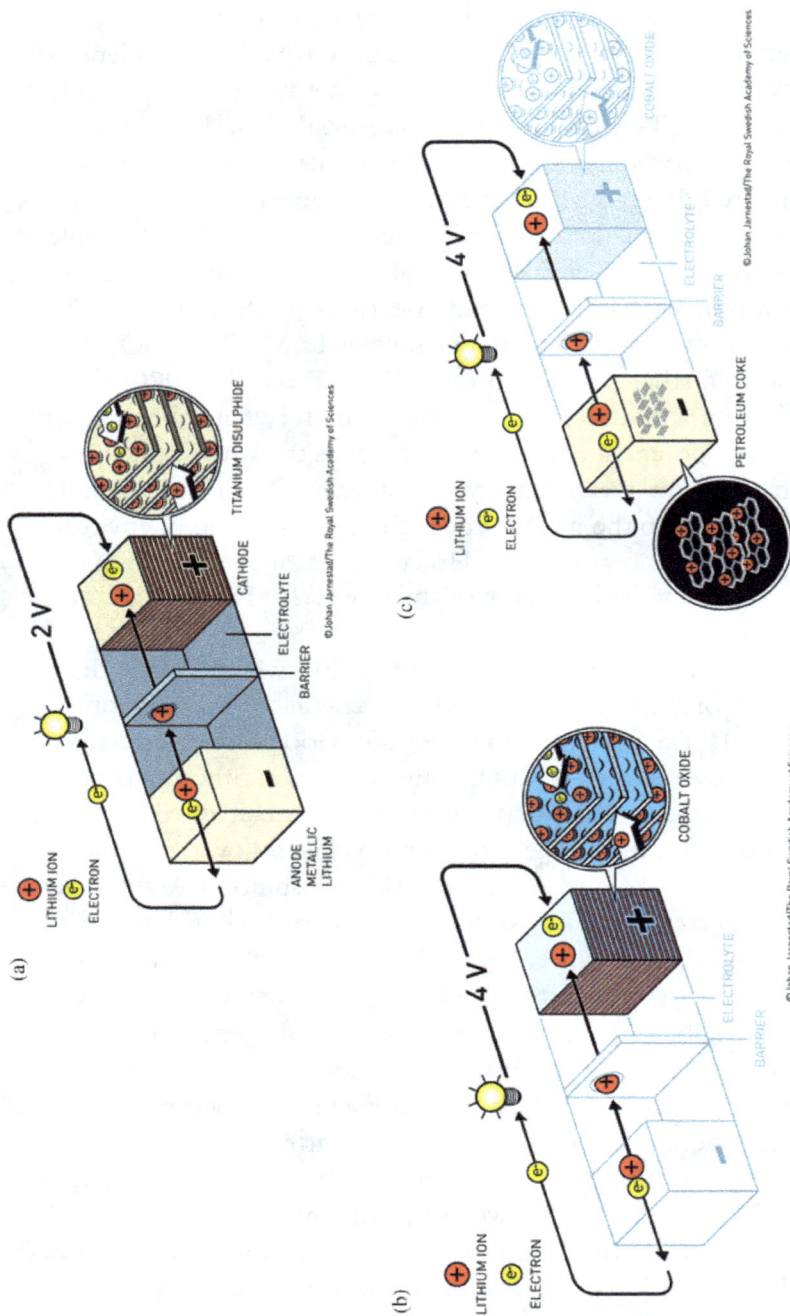

Fig. 1. The configurations of LRBs. (a) Whittingham's battery comprising a Li anode and TiS$_2$ cathode with a potential of ~2 V. (b) Goodenough's battery consisting of a Li anode and LiCoO$_2$ cathode. The use of cobalt oxide in the battery's cathode almost doubled the battery's potential. (c) Yoshino's battery based on a LiCoO$_2$ cathode and petroleum coke anode. This is the first commercially viable LRB. Copyright © Johan Jarnestad/The Royal Swedish Academy of Sciences. Reproduced with permission.

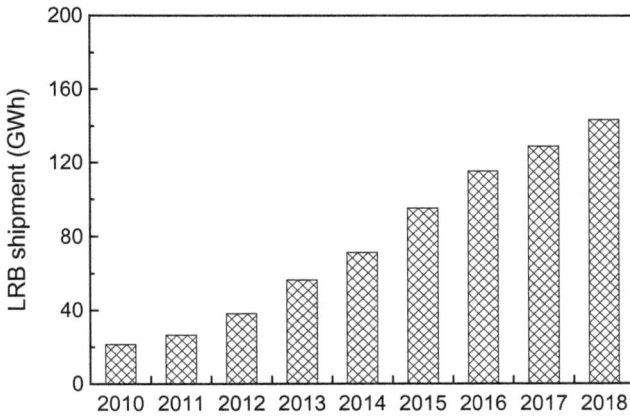

Fig. 2. Shipment of LRBs from 2010 to 2018. Data are collected from SPIR.

and revolutionized our daily lives. Today, LRBs are used in everything from cell phones to laptops and electric automobiles. Moreover, LRBs have also enabled the development of energy storage systems for smart grids and renewable sources. The rise in demand for consumer electronic devices along with the advent of EVs and HEVs propels the development of the LRB market, which has witnessed continuous growth during the last three decades.[9] A statistical report from SPIR shows that the shipment of LRBs is about 21.5 GWh in 2010, and drastically increases to 95.2 GWh in 2015 (Fig. 2). In 2018, the LRBs market amounts to 143.5 GWh with a revenue of $44.2 billion, which is about seven-fold that of seven years ago. Yet the majority of LRBs are applied in consumer electronic devices; the market for automotive and stationary storage applications are growing swiftly. Deployment in these areas demands LRBs with higher energy and power with less cost, which requests further advancement of current LRB technologies.

2. Basic Battery Concepts

Strictly speaking, a battery consists of two or more cells connected in series or parallel, but the term is also used for a single cell. A battery transforms chemical energy into electrical energy through electrochemical reactions occurring at its electrodes. In practice, the terms anode and cathode are used to represent the negative and positive electrodes, respectively for a wealth of batteries. In this section, the battery classification, working principle, and battery characteristics will be included as basic battery concepts.

2.1. *Battery classification*

Electrochemical cells and batteries can be divided depending on their component, shape, characteristics, application, etc. For example, batteries can be classified as liquid battery, polymer battery, and solid-state battery according to the electrolytes used. Batteries can also be divided into prismatic, cylindrical, and coin cells by the shape. With respect to the operating temperature, high-temperature and room-temperature cells can be defined. In principle, the battery can be divided into chemical battery, physical battery, and bio-battery. Chemical batteries, in which electrical energy is produced by conversion of chemical energy via redox reactions at the anode and cathode, can be further subdivided into primary or rechargeable batteries, depending on their rechargeability.[1]

Primary batteries are not capable of being easily or effectively recharged electrically. They are most commonly used in portable devices that have a low current drain, are used only intermittently, or are used well away from an alternative power source. Lithium primary batteries that generally use metallic lithium as the anode are a convenient, inexpensive, and lightweight power source for portable electronics. The general advantages of lithium primary batteries are good shelf life, high energy density at low to moderate discharge rates, and ease of use. Specifically, they are attractive for military applications due to high energy densities.

Rechargeable batteries can be recharged by applying electric current, which reverses the chemical reactions that occur during discharge. LRBs have a high initial cost but can be recharged very cheaply and used many times. Also, they have a lower environmental impact than lithium primary batteries. Therefore, they are widely used for portable consumer devices, power tools, and uninterruptible power supplies. Emerging applications in automobiles are driving the technology to reduce cost and weight and increase lifetime. Another important application is in grid energy storage applications for load leveling, where they store electric energy for use during peak load periods, and for renewable energy uses, such as storing power generated from photovoltaic arrays during the day to be used at night.

2.2. *Working principle*

Typically, LRBs consist of Li-containing cathode such as $LiCoO_2$, $LiMn_2O_4$, or $LiFePO_4$, a carbonaceous anode like graphite, and an electrolyte solution containing salt such as $LiPF_6$.[10,11] A typical $LiCoO_2//C$ rechargeable battery system can be written as follows:

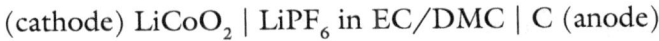

(cathode) $LiCoO_2$ | $LiPF_6$ in EC/DMC | C (anode)

Taking this $LiCoO_2//C$ battery as an example, the reaction mechanism of LRBs is illustrated in Fig. 3. The positive and negative materials usually have a layered structure to facilitate the intercalation of Li^+ ions. The net effect of the charge and discharge reaction is the movement of Li^+ ions back and forth between the electrodes, matched by a corresponding flow of electrons in the external circuit. During charge, Li^+ ions are extracted from the van der Waals gap between the O layers in the $LiCoO_2$ lattice, and intercalate into graphene interlayers in the graphite anode, accompanied by movement of electrons from the cathode to the anode. The charge balance is maintained by oxidation of Co^{3+} to Co^{4+}. Upon discharge, Li^+ ions flow from the anode and return to the cathode, and the electrons travel through the connection to the cathode and combine with the lithium ions, accompanied with a reduction of the Co^{4+} ions to Co^{3+}. During the electrochemical process, the cathode reaction is $LiCoO_2 \leftrightarrow Li_{1-x}CoO_2 + xLi^+ + xe^-$, while the anode reaction is $6C + xLi^+ + xe^- \leftrightarrow Li_xC_6$. The overall reaction can be written as $LiCoO_2 + 6C \leftrightarrow Li_{1-x}CoO_2 + Li_xC_6$. Note that the electrode process is determined by the reaction direction. From right to left is the discharge process, while from left to right is the recharge process.

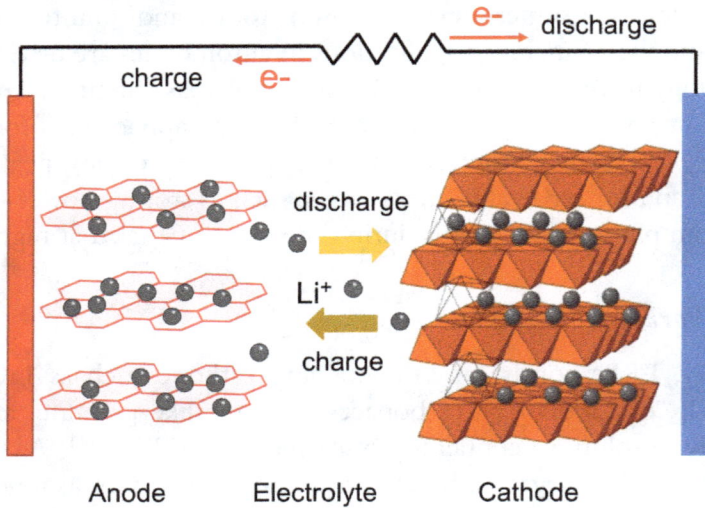

Fig. 3. Principle of an LRB based on LiCoO$_2$ cathode and carbon anode (graphite).

Figure 4 shows a schematic energy diagram of a cell at open circuit, where the redox energies of the cathode (E_c) and anode (E_a) should lie within the bandgap E_g of the electrolyte to dismiss electrolyte participation in electrochemical redox process.[12] An anode with a $\mu_{a(Li)}$ higher than the LUMO will reduce the electrolyte; likewise, a cathode with a $\mu_{c(Li)}$ less than the HOMO will oxidize the electrolyte. The operation potentials can be expanded by the formation of solid electrolyte interphase (SEI) on the surface of the electrode. SEI passivation layer is an ionic conductor but an electronic insulator. Therefore it can block electron transfer from the electrolyte HOMO to the cathode, or from anode to electrolyte LUMO. Thus, a battery with electrode redox energies (E_c or E_a) with a certain boundary outside of the "stable window" is also possible.

If cathode and anode are known, the open-circuit voltage can be calculated from $V_{oc} = E_c - E_a$, where E_c and E_a are the cathode and anode potential, respectively. For a typical LiCoO$_2$//graphite LRB, the V_{oc} is 3.7 V. The open-circuit voltage, capacity, and specific energy of typical LBR systems are listed in Table 1.

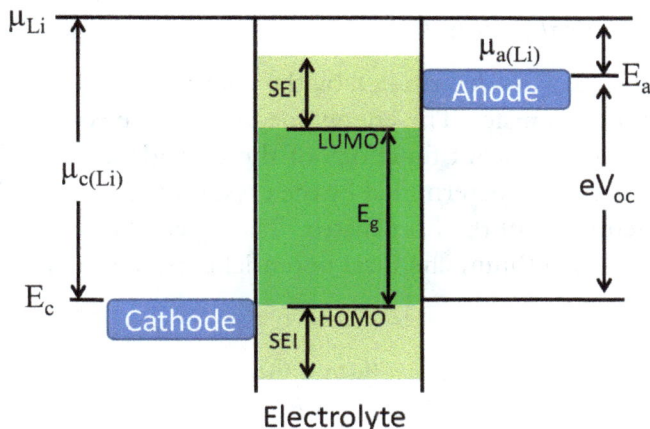

Fig. 4. Schematic energy diagram of a cell at open circuit. $\mu_{c(Li)}$ and $\mu_{a(Li)}$ are the Li chemical potential in cathode and anode, respectively. E_g is the thermodynamically stable window of the electrolyte. E_c and E_a are the redox energies of the cathode and anode vs. Li. LUMO and HOMO are referred to as the lowest unoccupied molecular orbital and the highest occupied molecular orbital of the electrolyte. SEI is the solid/electrolyte interphase. Adapted with permission.[12] Copyright ACS 2010.

Table 1. Electrochemical characteristics of typical LBR systems.

Cathode	$V_{oc}(V)$[a]	Cathode capacity (mAh g^{-1})[b]	Specific energy (Wh kg^{-1})[b]
LiCoO$_2$//graphite	3.7	140	518
LiNiO$_2$//graphite	3.6	150	540
LiMn$_2$O$_4$//graphite	3.8	120	464
LiNi$_{1-x-y}$Co$_x$Mn$_y$O$_2$//graphite	3.6	160	576
LiFePO$_4$//graphite	3.2	160	512
LiCoO$_2$//Li$_4$Ti$_5$O$_{12}$	2.3	140	322

[a] Note that the V_{oc} depends on the battery's state of charge. Here is the average V_{oc}.

2.3. *Battery characteristics*

A battery has various characteristics that define the type and capacity of a battery. The main characteristics of a battery are given below.

2.3.1. *Theoretical potential*

The cell voltage V_{oc} is determined by the energies involved in electron transfer and Li$^+$ transfer. The energy magnitude of electron transfer is associated with the work functions of the cathode and anode, while that of Li$^+$ transfer is determined by the crystal structure and the coordination geometry of the host matrix. The V_{oc} can be obtained by the difference in the lithium chemical potential between the cathode and the anode as

$$V_{oc} = \frac{\mu_{c(Li)} - \mu_{a(Li)}}{nF},$$ [1]

where $\mu_{c(Li)}$ is Li chemical potential of the cathode, $\mu_{a(Li)}$ is Li chemical potential of the anode, n is the number of transferable electrons and F, the Faraday constant (96,487 Coulombs).

The chemical potential a material can be expressed by its activity, α_i, as

$$\mu_i = \mu^o + RT \ln \alpha_i,$$ [2]

where μ_i is the chemical potential of relevant species, α_i, activity, R, gas constant, and T, absolute temperature. Therefore open-circuit voltage can also be written as a function of activity:

$$V_{oc} = \left(\mu^o_{c(Li)} - \mu^o_{a(Li)}\right) - RT\left[\frac{1}{F}\ln\left(\frac{\alpha_c}{\alpha_a}\right)\right],$$ [3]

where $\mu^o_{c(Li)} - \mu^o_{a(Li)}$ is the potential difference at standard condition.

2.3.2. *Internal resistance*

Generally, not all chemical energy can be converted to electric energy during the battery discharge process. The loss of the energy is due to polarization when a load current passes through the electrodes. The polarization is caused by two factors, one of which is Ohmic resistance R_Ω. Due to the existence of internal resistance, following Ohm's law, there exists a linear relationship between resistance and voltage drop at a given current. R_i is the sum of the electronic resistances of the active mass, the ionic resistance of the electrolyte, the current collectors and separator, and the contact resistance between the active mass and the current collector. Another one is polarization resistance R_p, which includes electrochemical activation resistance on both electrodes and concentration polarization. The former drives the electrochemical reaction at the electrode surface and the latter arises from the concentration differences of the reactants and products at the electrode surface and in the bulk as a result of mass transfer.

2.3.3. *Working voltage*

Working voltage, U_{wp}, is defined as the potential difference between the cathode and anode as current passes through electrodes. U_{wp} is always below potential difference, E, of a cell due to internal resistance:

$$U_{wp} = E - \eta^+ - \eta^- - IR_\Omega, \qquad [4]$$

where η^+ is the overpotential in cathode, η^- the overpotential in anode, I the past current, and R_Ω the Ohmic resistance. In principle, U_{wp} is associated with battery discharge protocol in terms of discharge time, discharge current, ambient temperature, cut-off voltage, etc.

2.3.4. *Capacity and specific capacity*

A battery's capacity is the amount of electric charge that can be stored in the material. In a real cell, the capacity is determined by the amount of active materials. The more active the material, the greater the capacity of the cell. A battery's capacity C can be calculated as

$$C = xnF, \qquad [5]$$

where x is the molar number of active material involved in the redox reaction, n, charge transfer number per mole, F, the Faraday constant. Note that the capacity of a battery depends on the discharge conditions such as the current rate, the cut-off voltage, the temperature, and other factors.

Specific capacity C_s is the capacity per unit mass of a cell or active material. C_s can be computed by

$$C_s = \frac{nF}{M},$$ [6]

where M is the molar weight of active material. Since one mole electron transfer will deliver 96,487 C or 26.8 Ah, the specific capacity of a battery or active material based on the term of Ah g^{-1} can be rewritten as

$$C_s = \frac{26.8n}{M}.$$ [7]

2.3.5. *Energy and specific energy*

Energy is the work that can be delivered by a specific electrochemical system. The energy generally in Wh is evaluated by taking both the voltage and the quantity of electricity into consideration. The energy can be calculated as

$$W = \int U_{wp} dC.$$ [8]

If the discharge plateau is relatively flat, an average value of voltage is often used:

$$W = CU_{av},$$ [9]

where U_{av} is the average voltage during discharge.

Specific energy W_s is used in place of gravimetric energy density, defined as energy per unit mass:

$$W_s = \frac{CU_{av}}{m}.$$ [10]

2.3.6. *Power and specific power*

Power (P in Watt) is the work that can be delivered by a specific electrochemical system in a certain time duration:

$$P = \frac{W}{t} = \frac{CU}{t} = \frac{ItE}{t} = IE.$$ [11]

Specific power (P_s in Watt kg^{-1}) is used in place of gravimetric power density, defined as power per unit mass:

$$P_s = \frac{W}{tm} = \frac{CU}{tm} = \frac{ItE}{tm} = \frac{IE}{m}. \tag{12}$$

2.3.7. Depth of discharge

Depth of discharge (DOD) is defined as a percentage of the nominal capacity; 0% DOD means no discharge, 50% DOD means half discharge, and 100% DOD means full discharge. Due to variations during manufacture and aging, the DOD for complete discharge can change over time or number of charge cycles. Normally, a rechargeable battery system can tolerate more charge/discharge cycles if the DOD is lower on each cycle.

3. Materials Overview

The performance of LRBs depends heavily on the types of materials and processing technologies. Since there is a risk that LRBs can undergo thermal runaway, materials play an important role in securing safe performance. With an increase in the scale of energy storage such as in EV and HEV, safety associated with electrode materials becomes very crucial and important recently.[13] Hence, a brief overview covering major materials systems in LRBs including cathode, anode, electrolyte, and separator is highly desirable in this chapter.

3.1. Cathode materials

The intercalation compounds are among the most valuable cathode materials. In these compounds, a guest species such as lithium can be inserted interstitially into the host lattice and subsequently extracted during recharge with little or no structural modification of the host. Key requirements for cathode material include (1) high free energy of reaction with lithium, (2) wide range of Li$^+$ ion intercalation, (3) little structural change on reaction, (4) highly reversible reaction, (5) rapid

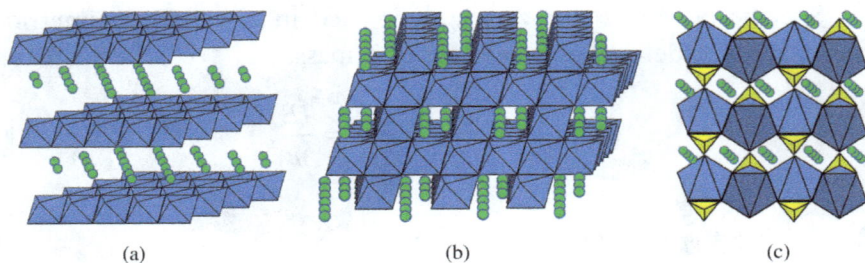

Fig. 5. Structures of some typical cathode materials. (a) Layered $LiCoO_2$, (b) spinel $LiMn_2O_4$, and (c) olivine $LiFePO_4$. The blue octahedral signals MO_6 groups, yellow tetrahedral PO_4 groups, green circle Li^+ ions.

diffusion of Li^+ ion in solid, (6) good electronic conductivity, (7) no solubility in electrolyte, and (8) low cost and environmental benignity.[7]

This implies that the first-line 3d transition metal chalcogenides should be the preferable choice. Currently, lithiated transition metal oxides (such as $LiCoO_2$, $LiMn_2O_4$) and phosphates have been broadly exploited in LRBs. Figure 5 schematically presents the crystalline structures of these commercially exploited cathode materials including layered $LiCoO_2$, spinel $LiMn_2O_4$, and olivine $LiFePO_4$. Their electrochemical characteristics are summarized in Table 2 and Fig. 6.

3.1.1. *Layered LiCoO₂*

$LiCoO_2$ has a layered structure, where Li and Co cations occupy alternative layers of octahedral sites in a distorted cubic close-packed O lattice. The layered metal oxide framework provides a two-dimensional interstitial space, which allows for each removal of the Li^+ ions.[14] However, only ~0.5 Li/Co can be reversibly removed and inserted from $LiCoO_2$ without causing cell capacity loss due to changes in the $LiCoO_2$, leading to a capacity of 140 mAh g^{-1}. By modifying $LiCoO_2$ with metal oxides such as Al_2O_3, ZrO_2, and TiO_2, a reversible and stable capacity above 170 mAh g^{-1} can be obtained.[15] Cho *et al.* attributed the improved capacity and cyclability to suppression of both the changes in the c axis and hexagonal to monoclinic phase transition. In contrast, Dahn *et al.* proposed surface protection for

Table 2. Electrochemical characteristics of representative cathode materials for LBRs.

Cathode	$LiCoO_2$	NMC	$Li_{1+x}M_{1-x}O_2$	$LiMn_2O_4$	$LiFePO_4$
Potential (V)	3.9	3.8	3.4	4.0	3.4
Theoretical capacity (mAh g^{-1})	274	~270	~300	148	170
Practical capacity (mAh g^{-1})	140 (4.3 V) 170 (4.5 V)	140 (4.3 V) 180 (4.5 V)	~250 (4.6 V)	120	160
Initial efficiency	95%	90%	~80%	95%	95%
Conductivity (S m^{-1})	1	10^{-3}	10^{-4}	10^{-3}	10^{-5}
Abundance	Scarce	medium	Medium	High	High
Cost	High	Medium	Medium	Low	Low
Safety	Poor	Good	Good	Good	Excellent

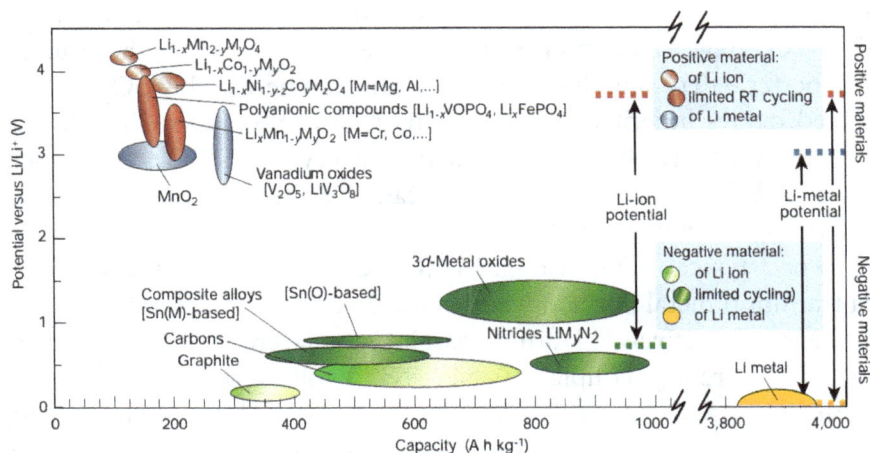

Fig. 6. Voltage versus capacity for cathode and anode materials currently used or developed for the next generation of LRBs. Reproduced with permission.[11] Copyright 2001, Springer-Nature.

such an enhancement in $LiCoO_2$.[16] The exact machismo is still under debate. Recently, Li and coworkers reported stable cycling of $LiCoO_2$ at 4.6 V through trace Ti–Mg–Al co-doping.[17] They revealed suppression of phase transition of $LiCoO_2$ at high voltages by Mg and Al incorporation, and stabilization of the surface oxygen by Ti segregating at grain boundaries and on the surface. This finding suggests a collaborating action of both surface and lattice.

Batteries produced with $LiCoO_2$ cathodes, while providing good capacity, are reactive and have poor thermal stability. This makes $LiCoO_2$ batteries more susceptible to thermal runaway in cases of abuse such as high-temperature operation or overcharging. At elevated temperatures, $LiCoO_2$ decomposes and releases oxygen, which then reacts exothermically with the organic materials in the cell, and finally induces thermal runaway in the battery systems. Therefore, $LiCoO_2$-based LRBs can only be adopted in small size.

3.1.2. *Layered* $LiNi_{1-x-y}Co_xMn_yO_2$

Ternary layered oxides have been used as positive electrodes. Two ternary electrode systems, $LiNi_{1-x-y}Co_xMn_yO_2$ (NMC) and $LiNi_{1-x-y}Co_xAl_yO_2$ (NMA) are intensively exploited due to their higher capacity, better safety and less toxicity than commercial $LiCoO_2$.[18] The mixed oxides inherit the merits of mono-metal oxide such as good cyclability of $LiCoO_2$, high capacity delivery of $LiNiO_2$, and super safety of $LiMnO_2$, and these features can be elaborately tuned through the ratio of metal elements.

For a typical $LiNi_{1/3}Co_{1/3}Mn_{1/3}O_2$ ternary material, the valences of transition metal in the mixed oxides are 2+, 3+, and 4+ for Ni, Co, and Mn, respectively.[19] During charging, the layered materials involve firstly the redox couple of Ni^{2+}/Ni^{4+}, followed by Co^{3+}/Co^{4+}. $LiNi_{1/3}Co_{1/3}Mn_{1/3}O_2$ exhibits a capacity of 150 mAh g^{-1} when charged to 4.3 V, and delivers a much higher capacity of ~200 mAh g^{-1} if the cut-off voltage is raised to 4.5 V or above. Cycling at a high cut-off voltage does not result in an irreversible structural transition but may lead to irreversible capacity loss due to highly active Co^{4+} species. Therefore, coating with inert oxides such as Al_2O_3, ZrO_2, TiO_2 was generally adopted to achieve high capacity and stable cycling of NMC. The coating layer can prevent the direct contact of the electrode from the electrolyte, and thus suppress the undesirable side reactions and dissolution loss of transition metal ions.[20–22]

In 2006, a lithium-excess layered material, $Li_{1+x}M_{1-x}O_2$, has been exploited. In this material, some Li occupies the transition metal site other than the Van der Waals gap, leading to a high capacity beyond 280 mAh g^{-1}.[23] However, Li-rich materials suffer from poor kinetics and evident voltage decay upon cycling, the reasons of which have not been well understood. Recently, by using operando three-dimensional Bragg coherent diffractive imaging, Shpyrko *et al.* could decipher the decay of voltage decay and establish a process to recover it.[24]

3.1.3. *Spinel LiMn$_2$O$_4$*

Spinel $LiMn_2O_4$ crystallizes in a three-dimensional structure via face sharing octahedral and tetrahedral. In the $LiMn_2O_4$ structure, the O ions form a cubic close-packed array, with Mn occupying half of the octahedral, and Li an eighth of the tetrahedral sites referring to the 16d and 8a sites, respectively. Therefore, its structure can be written as $[Li]_{8a}[Mn_2]_{16d}O_4$.[25] The edge-shared MnO_6 octahedra form a three-dimensional host, which provides conducting pathways for the insertion and extraction of lithium ions. About 0.8 Li can be reversibly extracted/inserted from/into $LiMn_2O_4$ host, resulting in a capacity of ~120 mAh g^{-1}.

The application of spinel $LiMn_2O_4$ materials is mainly hindered by its poor cycling stability, especially at temperatures above 50°C. Capacity fading has been concluded due to several factors, in which the Jahn-Teller distortion and Mn dissolution into electrolytes are regarded as the major ones.[26] The Jahn-Teller distortion associated with the high spin Mn^{3+} is as follows: d^4 ions will drive the cubic structure to tetragonal one with significant elongation in Z-axis. This structure transition results in a 16% increase in the *c/a* ratio and a 6.5% increase in unit-cell volume. The distortion is so severe that the crystalline integrity is destroyed, leading to a rapid capacity decay. Substitution of Mn by low-valence ions increases the average valence of Mn above +3.5, and thus significantly suppresses the Jahn-Teller distortion.[27] On the other hand, Mn dissolution due to the disproportionation of Mn^{3+} to MnO_2 and soluble Mn^{2+} is also one of the main reasons causing poor cycling. Surface modification with other metal oxides could prevent the $LiMn_2O_4$ from direct contact with the electrolyte and therefore suppress the dissolution of Mn species.[28]

3.1.4. *Olivine LiFePO$_4$*

Many metal oxide electrodes may experience oxygen release under overcharge or short circuit conditions. To hold the oxygen tightly in the lattice, a large number of polyanion structures have been explored, with $LiFePO_4$ being the most promising example. The strong covalent P–O bond not only enhances the Fe^{2+}/Fe^{3+} redox potential through an inductive effect but also prevents the oxygen atoms from releasing even in abused condition.[29] $LiFePO_4$ has a theoretical capacity of 170 mAh g^{-1}, with accessible one up to 160 mAh g^{-1} in practice and shows excellent cycling stability.

However, this phosphate suffers from poor Li (de)intercalation kinetics due to low electronic conductivity ($\sim 10^{-8}$ S cm^{-1}) and Li$^+$ ion diffusion ($\sim 10^{-14}$ cm^2 s^{-1}).[30] These limitations have been overcome to a large degree by carbon modification,[31] ion doping,[32] or nanostructuring.[33] $LiFePO_4$ is mainly regarded for automobile and large-scale energy applications due to the high safety level and low cost. Another two phosphates, namely $LiMnPO_4$ and $LiCoPO_4$, have been explored aiming at enhancing the specific energy of phosphate cathodes. Both materials share the same olivine structure as $LiFePO_4$ but exhibit a higher working potential, which is benefited for the augment in the voltage of full cells. In particular, the practice of $LiCoPO_4$ cathodes enables the design of $LiCoPO_4//Li_4Ti_5O_{12}$ batteries with a working voltage of 3.25 V, equal to that of a $LiFePO_4$/graphite cell.[34] However, the poor Li reaction kinetics[35,36] and lack of stable electrolyte working beyond 5 V[37,38] have limited their potential in practice.

3.2. *Anode materials*

Both cathode and anode materials impact battery performance in terms of energy density, power output, thermal stability, and cycle life. However, the safety and cycle life of LRBs are more dependent on the properties of anode materials. As the redox energy of graphite, which is now the prevailing anode applied in most LRBs, is above the LUMO of electrolyte, electrolyte reduction can only be kinetically blocked by an SEI passivation layer. Therefore, the SEI film on the anode will significantly impact the cycling stability of LRBs. In addition, lithiated graphite (Li_xC_6) is highly reactive and may catch fire when the battery is overcharged or heated to high temperatures. In this section, I will briefly discuss some important anode materials for LRBs, focusing on their Li-storage mechanism and commercial perspective.

3.2.1. *Carbonaceous materials*

At present, carbon materials are widely used as the anode in LRBs since the chemical potential of lithiated carbon materials is close to that of metallic lithium. Thus an electrochemical cell made with a lithiated carbon material will have almost the same open-circuit voltage as one made with metallic lithium. The specific capacity, cyclic efficiency, and irreversible capacity loss of LRBs depend largely on the type of carbon materials used. Various types of carbonaceous materials have been investigated as the potential anode materials; they distinguish each other by their microstructure, texture, crystallinity, and morphology.[39,40]

Two typical types of carbon materials, ordered graphite and non-graphitizable hard carbon, are used in the anode for commercial batteries. The graphite anode includes natural and artificial graphite. While hard carbons were produced from cokes, polymers, fibers, and other precursors at a relatively low temperature about 1100°C,[41] fabrication of artificial graphite often needs a much higher temperature (>2500°C) to convert the carbon to ordered graphite layers. The graphite electrode can deliver a reversible capacity of 300–350 mAh g⁻¹, with a lithium insertion/extraction plateau below 0.2 V. In the first cycle, an irreversible capacity loss accounting for 8–10% of the total capacity occurs at about 0.8 V due to electrolyte decomposition and formation of SEI film.[42,43] During subsequent cycles, the irreversible capacity is much reduced, and the electrode exhibits stable cycling. The merits of large capacity, low voltage plateau, stable cycling behavior, and low cost make graphite materials particularly suitable for the LRB industry. Generally, artificial graphite outperforms natural graphite but the latter is cheap and adequate in terms of energy density.

On the other hand, hard carbon offers a higher theoretical capacity up to 500–700 mAh g^{-1} due to its highly irregular and disordered structure.[44] This structure consists of a significant amount of single-layer carbon atoms that can adsorb Li$^+$ ions in both sides of the carbon graphene sheet. Moreover, the space gap between the adjacent carbon layers is larger than graphite materials, which is beneficial for Li$^+$ ion mobility.[45] As a result, hard carbon displays a high level of power, which is particularly desirable for power sources used in automobiles and power tools. However, hard carbon materials also exhibit some drawbacks, including large irreversible capacity, low pack density, and hysteresis in the voltage profile. These drawbacks are intimately related to the physical and chemical characteristics of hard carbon like stacking fashion, pore size and distribution, and impurities. Therefore, modifications to address these issues to manipulate the capacity delivery and efficiency are highly necessary.[46]

3.2.2. Lithium titanate

Some LRBs use lithium titanate ($Li_4Ti_5O_{12}$) as the anode, most notably Toshiba's SCiB. $Li_4Ti_5O_{12}$ exhibits a higher redox potential of ~1.55 V, which can avoid the reduction of the electrolyte on the electrode surface and the formation of a solid electrolyte interface.[47] Additionally, the zero-strain characteristic upon Li insertion endows it excellent reversibility, structural stability, and Li$^+$ ion mobility.[48] The major challenge facing its application is the poor conductivity (ca. 10^{-13} S cm^{-1}) associated with a bandgap energy of ~2 eV due to empty Ti 3d-states. Therefore, $Li_4Ti_5O_{12}$ generally needs special engineering such as surface coating and/or size reduction to meet the high current output demand.[49,50] At present, this material is increasingly used in automotive and storage applications where safety, product lifespan, and cost are of prime importance.

3.2.3. *Silicon*

Silicon materials are particularly attractive anodes for next-generation LRBs due to their extremely high capacity of ~4200 mAh g^{-1}, as each silicon atom can accommodate 4.4 lithium atoms leading to the formation of Li$_{4.4}$Si alloy. Also, Si displays a very flat voltage plateau at 0.2 V during Li alloying, similar to a graphite anode. Moreover, silicon is the second most abundant element on earth. Therefore, much attention has been focused on developing silicon anode for LRBs.[51-53] However, the alloying process of Li with Si results in a three-fold expansion in volume, bringing significant particle pulverization and capacity fading. To mitigate this volume swelling, Si/C hybrid anodes have been proposed and gradually taken the commercial stage of LRBs.[54] Mitsubishi Chemical developed a Si/C composite, which exhibits a ~40% increase in capacity versus graphite.

3.2.4. *Metal oxides and sulfides*

Recently, transition metal oxides and sulfides emerge as a key type of electrode material for rechargeable batteries.[55,56] These materials adopt a distinct conversion mechanism to store a large amount of Li$^+$ ions, in contrast to the typical intercalation mechanism occurring in graphite.[57] During the discharge process, the oxides (or sulfides) are reduced to metal, accompanied with the formation of Li$_2$O (Li$_2$S), and vice versa upon recharge process. Hence, the total process can be written as MeO$_x$ + 2xLi $\leftrightarrow$ Me + xLi$_2$O. Due to poor electrochemical activity, the conversion-type anodes have often been engineered into nanostructures[58] or composites with nanocarbon.[59] Despite a high capacity, large irreversible loss and significant voltage hysteresis make them unlikely to be used in the commercial LRBs.[60]

3.3. *Electrolyte*

The choice of electrolytes for LRBs is also critical. The electrolyte should have the characteristics including but not limited to (a) good ionic conductivity (10^{-3} S cm^{-1}), (b) unity of Li$^+$ ion transference number, (3) wide electrochemical voltage window (0 to 5 V), (4) thermal stability (up to 70°C), and (5) compatibility with other cell components.[61]

Currently, non-aqueous liquid and polymer electrolytes are two main types for LRBs. Liquid electrolytes adopt aprotic organic electrolyte solvents, such as propylene carbonate, ethylene carbonate, diethyl carbonate, and ethyl methyl carbonate because of their low reactivity with lithium. These organic liquid electrolytes generally have conductivities that are about two orders of magnitude lower than aqueous electrolytes. To improve the battery performance, electrolyte additives such as vinylene carbonate, propane sultone are adopted to facilitate the formation of SEI membrane on the surface of graphite, reduce irreversible capacity and gas generation, and stabilize long-term cycling.[62,63]

An alternative to the liquid electrolytes is a solid polymer electrolyte formed by incorporating lithium salts into polymer matrices and casting into thin films. These can be used as both the electrolyte and separator. These electrolytes have lower ionic conductivities and low lithium-ion transport numbers compared to the liquid electrolytes, but they are less reactive with lithium, which should enhance the safety of the battery. The use of thin polymer films or operation at higher temperatures (60–100°C) compensate in part for the lower conductivity of the polymer film. The solid polymers also offer the advantages of a "nonliquid" battery and the flexibility of designing thin batteries in a variety of configurations.

3.4. *Separators*

Separators are permeable films placed between a battery's anode and cathode with a number of pores. The main function of a separator is to separate the electrodes to electronically isolate both electrodes while also allowing the transport of ionic charge carriers. Separator materials tend to be polyolefins like polyethylene or polypropylene.[64]

Separators are critical components in liquid electrolyte batteries. A separator generally consists of a polymeric membrane forming a microporous layer. It must be chemically and electrochemically stable concerning the electrolyte and electrode materials and mechanically strong enough to withstand the high tension during battery construction. They are important to batteries because their structure and properties considerably impact the battery performance, including the batteries energy and power densities, cycle life, and safety.[65]

Separators for LRBs include polymer films such as polyethylene, polypropylene, polyvinyl chloride. These polymer separators can be fabricated by either a dry or a wet process. The wet process includes the stretch of polymer materials at the assistance of solvents. Separators produced via the wet process show a high physical strength and robust thermal shutdown function.

In the dry process, polymers are mechanically stretched without using solvents; therefore, the production cost can be much reduced. Separators made by the dry process generally consist of pores with high permeability but poor uniformity in the size. Currently, separators by the wet process are used for consumer electronics, while separators by the dry process are used for automotive batteries and low-end batteries for consumer electronics.

4. Battery Design and Manufacture

4.1. *Battery design*

Although the performance of LRBs principally depends on the basic electrochemical reactions at both electrodes, the battery design also has a substantial impact on the magnitude of the energy, power, efficiency, and calendar life. Thus, some essential requirements should be obeyed in the design of batteries to achieve high operating efficiency with minimal loss of energy.

The successful battery design is a definition of the key electrical parameters. These parameters are very important in choosing the correct chemistry and safety devices within the battery design. Operating voltage is a key component of this step. For LRBs, the upper voltage limit on charging is usually limited to 4.2 or 4.3 V, while the cell voltage should not fall below 2.5 volts.

The second key parameter needed for adequate battery design is the maximum current requirement. This is an important factor and will influence the battery designer's choice of protection circuitry, chemistry, wire and trace sizes and battery capacity. It is important to remember that the current used to design a system on a bench top using a power supply may be different than the current required from a battery.

The next step is to calculate the desired battery capacity based on the above requirements and the device runtime requirements. This will allow the proper-sized battery to be designed for the application. Remember that rechargeable systems will lose capacity as cycles of use accumulate. The desired runtime near the end of the useful life of a device should be considered and the battery appropriately sized to deliver the proper capacity at the later stages of device life.

As for the battery materials selection, the cathode must be an efficient oxidizing agent, be stable when in contact with the electrolyte, and have a useful working voltage. The electrolyte must have the following qualities: good ionic conductivity but not be electronically conductive, nonreactive to the electrode materials, little change in properties with the change in temperature, safety in handling, and low cost. The anode should be an efficient reducing agent with high Coulombic efficiency, good conductivity, stability, ease of fabrication, and low cost.

4.2. *Battery manufacture*

LRBs are manufactured in two phases: electrode fabrication and battery assembly. In the electrode process, cathode or anode material, as well as binders and conductive additives, are thoroughly mixed to form a uniform slurry. The slurry is then cast onto aluminum foil (cathode) and copper foil (anode) as the current collector, and dried and compressed using a roll press. Finally, the electrode is sliced into an appropriate size using a slitter and completed by winding. During the following assembly process, the positive and negative electrode sheets are sandwiched by the separator, and wound and inserted into the battery case. When electrolyte has been injected, the battery case is sealed using laser welding. The battery has been subjected to further inspection and formation before it is shipped to the market. It is noted that assembly processes may differ according to the shape of the battery, which can be cylindrical, prismatic, or laminated.

5. Summary

With the first commercial LRBs released into the market in 1991, almost 30 years have passed and rapid development has been witnessed. There is an accumulation of proliferation of different battery technologies and types, depending on the construction and materials used. With the world going green, the evolution of LRBs is open to innovations that will place it at the heart of the emerging EVs and smart electronics. Drastic changes in the LRB concept, structure, material, and electrochemistry are required. Improvements in energy density and environmental sustainability would be of high priority.

LRBs continuously evolve to satisfy increasing demands on both energy and power.[66,67] The prospect of using Li metal as the anode replacing graphite suggests that solid-state batteries may provide higher energy densities than the conventional LRBs. At present, many solid electrolytes show attractive properties such as high ion conductivity but negligible electronic one.[68] However, their interphases with the electrodes are of high resistance and restrain the power output of current solid-state batteries. In addition, there is growing research motion into $Li-O_2$ and Li–S batteries.[69,70] These systems work based on a conversion process rather than the intercalation, and thus can offer energy densities way beyond that of LRBs. However, technical challenges relative to electrolyte stability and charge-transfer kinetics have yet been well mitigated for these batteries.

Acknowledgments

The author is grateful to the financial support from the National Natural Science Foundation of China (Grant Nos. 51872192, 51672182), the Natural Science Foundation of Jiangsu Province (Grant No. BK20180002), the Natural Science Foundation of the

Jiangsu Higher Education Institutions of China (Grant No. 19KJA170001), and of the Priority Academic Program Development (PAPD) of Jiangsu Higher Education Institutions.

References

1. D. Linden and T. Reddy. *Handbook of Batteries*, 3rd edn. (McGraw Hill, USA, 2001).
2. M. Winter and R. J. Brodd, *Chem. Rev.* **104**, 4245 (2004).
3. G.-A. Nazri and G. Pistoia. *Lithium Batteries: Science and Technology.* (Springer, New York, 2009).
4. B. Dunn, H. Kamath and J. M. Tarascon, *Science* **334**, 928 (2011).
5. J. B. Goodenough, *Acc. Chem. Res.* **46**, 1053 (2012).
6. M. Winter, B. Barnett and K. Xu, *Chem. Rev.* **118**, 11433 (2018).
7. M. S. Whittingham, *Chem. Rev.* **104**, 4271 (2004).
8. *Scientific Background on the Nobel Prize in Chemistry 2019.* (The Royal Swedish Academy of Sciences, Sweden, 2019).
9. M. Li *et al.*, *Adv. Mater.* **30**, 1800561 (2018).
10. J. B. Goodenough, *Nat. Electron.* **1**, 204 (2018).
11. J. M. Tarascon and M. Armand, *Nature* **414**, 359 (2001).
12. J. B. Goodenough and Y. Kim, *Chem. Mater.* **22**, 587 (2010).
13. F. Cheng *et al.*, *Adv. Mater.* **23**, 1695 (2011).
14. K. Mizushima *et al.*, *Mater. Res. Bull.* **15**, 783 (1980).
15. J. Cho *et al.*, *Angew. Chem. Int. Ed.* **40**, 3367 (2001).
16. Z. Chen and J. R. Dahn, *Electrochem. Solid-State Lett.* **6**, A221 (2003).
17. J.-N. Zhang *et al.*, *Nat. Energy* **4**, 594 (2019).
18. A. Chakraborty *et al.*, *Chem. Mater.* **32**, 915 (2020).
19. A. Manthiram, B. Song and W. Li, *Energy Storage Mater.* **6**, 125 (2017).
20. S.-T. Myung *et al.*, *Chem. Mater.* **17**, 3695 (2005).
21. J. Ni *et al.*, *Electrochim. Acta* **53**, 3075 (2008).
22. Y. Huang *et al.*, *J. Power Sources* **188**, 538 (2009).
23. P. Rozier and J. M. Tarascon, *J. Electrochem. Soc.* **162**, A2490 (2015).
24. A. Singer *et al.*, *Nat. Energy* **3**, 641 (2018).
25. M. M. Thackeray *et al.*, *Mater. Res. Bull.* **18**, 461 (1983).
26. G. Amatucci and J. M. Tarascon, *J. Electrochem. Soc.* **149**, K31 (2002).
27. Z. Chen and K. Amine, *J. Electrochem. Soc.* **153**, A1279 (2006).
28. Y. K. Sun, K. J. Hong and J. Prakash, *J. Electrochem. Soc.* **150**, A970 (2003).

29. A. K. Padhi, K. S. Nanjundaswamy and J. B. Goodenough, *J. Electrochem. Soc.* **144**, 1188 (1997).
30. J. Ni *et al.*, *Carbon* **92**, 15 (2015).
31. Z. Chen and J. R. Dahn, *J. Electrochem. Soc.* **149**, A1184 (2002).
32. S.-Y. Chung, J. T. Bloking and Y.-M. Chiang, *Nat. Mater.* **1**, 123 (2002).
33. J. Ni *et al.*, *J. Power Sources* **195**, 2877 (2010).
34. J. Ni *et al.*, *Electrochem. Commun.* **35**, 1 (2013).
35. J. Ni and L. Gao, *J. Power Sources* **196**, 6498 (2011).
36. J. Ni *et al.*, *J. Power Sources* **196**, 8104 (2011).
37. J. Ni *et al.*, *Electrochim. Acta* **70**, 349 (2012).
38. J. Ni *et al.*, *Electrochem. Commun.* **31**, 84 (2013).
39. J. R. Dahn *et al.*, *Science* **270**, 590 (1995).
40. M. Endo *et al.*, *Carbon* **38**, 183 (2000).
41. I. E. Moctar *et al.*, *Funct. Mater. Lett.* **11**, 1830003 (2018).
42. M. Yoshio, H. Wang and K. Fukuda, *Angew. Chem. Int. Ed.* **42**, 4203 (2003).
43. K. Xu, *J. Electrochem. Soc.* **154**, A162 (2007).
44. J. R. Dahn, W. Xing and Y. Gao, *Carbon* **35**, 825 (1997).
45. J. Ni, Y. Huang and L. Gao, *J. Power Sources* **223**, 306 (2013).
46. H. Haruna *et al.*, *J. Power Sources* **196**, 7002 (2011).
47. J. Ni *et al.*, *J. Solid State Electrochem.* **16**, 2791 (2012).
48. L. Zhao *et al.*, *Adv. Mater.* **23**, 1385 (2011).
49. L. Cheng *et al.*, *J. Electrochem. Soc.* **154**, A692 (2007).
50. Y.-Q. Wang *et al.*, *J. Am. Chem. Soc.* **134**, 7874 (2012).
51. U. Kasavajjula, C. Wang and A. J. Appleby, *J. Power Sources* **163**, 1003 (2007).
52. A. Magasinski *et al.*, *Nat. Mater.* **9**, 353 (2010).
53. L.-F. Cui *et al.*, *Nano Lett.* **9**, 491 (2009).
54. M. L. Terranova *et al.*, *J. Power Sources* **246**, 167 (2014).
55. M. V. Reddy, G. V. Subba Rao and B. V. Chowdari, *Chem. Rev.* **113**, 5364 (2013).
56. J. Ni *et al.*, *Nano Energy* **34**, 356 (2017).
57. P. Poizot *et al.*, *Nature* **407**, 496 (2000).
58. P.-L. Taberna *et al.*, *Nat. Mater.* **5**, 567 (2006).
59. J. Ni and Y. Li, *Adv. Energy Mater.* **6**, 1600278 (2016).
60. Y. Cao *et al.*, *Nat. Nanotechnol.* **14**, 200 (2019).
61. D. Aurbach *et al.*, *Electrochim. Acta* **50**, 247 (2004).
62. S. S. Zhang, *J. Power Sources* **162**, 1379 (2006).

63. K. Xu, *Chem. Rev.* **114**, 11503 (2014).
64. P. Arora and Z. Zhang, *Chem. Rev.* **104**, 4419 (2004).
65. S. S. Zhang, *J. Power Sources* **164**, 351 (2007).
66. J. Ni, *Funct. Mater. Lett.* **11**, 1802001 (2018).
67. K. Chen and D. Xue, *Funct. Mater. Lett.* **12**, 1830005 (2019).
68. F. Zheng *et al.*, *J. Power Sources* **389**, 198 (2018).
69. J. Lu *et al.*, *Chem. Rev.* **114**, 5611 (2014).
70. M. Li *et al.*, *Adv. Mater.* **30**, 1801190 (2018).

Chapter 2

Carbon-Metal Oxide Nanocomposites as Lithium-Sulfur Battery Cathodes

Sheng Zhu* and Yan Li*,†

In rechargeable lithium-sulfur (Li-S) batteries, the conductive carbon materials with high surface areas can greatly enhance the electrical conductivity of sulfur cathode, and metal oxides can restrain the dissolution of lithium polysulfides within the electrolyte through strong chemical bindings. The rational design of carbon-metal oxide nanocomposite cathodes has been considered as an effective solution to increase the sulfur utilization and improve cycling performance of Li-S batteries. Here, we summarize the recent progresses in the carbon-metal oxide composites for Li-S battery cathodes. Some insights are also offered on the future directions of carbon-metal oxide hybrid cathodes for high performance Li-S batteries.

Keywords: Carbon; metal oxides; rechargeable; lithium-sulfur batteries; cathodes.

1. Introduction

Rechargeable lithium-sulfur (Li-S) batteries have received considerable interest since their discovery in the 1960s,[1] are low-cost and have

*Beijing National Laboratory for Molecular Sciences, Key Laboratory for the Physics and Chemistry of Nanodevices, State Key Laboratory of Rare Earth Materials, Chemistry and Applications, College of Chemistry and Molecular Engineering, Peking University, Beijing, P. R. China.
†yanli@pku.edu.cn

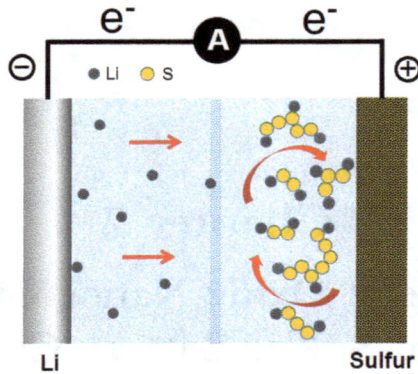

Fig. 1.   Schematic of the electrochemistry of Li-S batteries.

overwhelming advantages in high theoretical capacity (1672 mAh g^{-1}) and energy density (2600 Wh kg^{-1}) calculated by the multielectron transfer reaction: $S_8 + 16Li^+ + 16e^- \rightarrow 8Li_2S$.[2–5] A specific schematic illustration of main components in a Li-S battery and its charge/discharge process is displayed in Fig. 1. Upon discharging, Li ions and electrons are produced by the oxidization of lithium metal (negative electrode) and transfer to the S cathode (positive electrode) through electrolyte, resulting in reduction of S to produce lithium sulfide. From this point of view, there are several major challenges that hinder the practical applications for Li-S cells, such as: (1) the low electrochemical utilization and limited rate capability due to low electrical conductivity of sulfur and polysulfides. (2) The huge volume expansion (about 79%) during the full reduction of sulfur into Li$_2$S, and (3) the low capacity and poor cycling performance originated from the dissolution of lithium sulfide into the electrolyte because of their spontaneous redox reactions.[6–10]

Great efforts, including the development of advanced composite cathodes,[11–13] protected anodes,[14–16] and electrolytes[17–19] have been devoted to increasing the properties of Li-S cells. A promising approach to enhance the overall electric conductivity of the cathode and prevent the dissolution of lithium polysulfides during cycling is to restrict sulphur species into a conductive carbon matrix, such as: graphene,[20–23] carbon nanotube (CNT),[13,24,25] microporous–mesoporous carbon,[26,27] porous graphitic carbon,[28,29] porous hollow carbon,[30-33] hollow carbon nanosphere,[34,35] and so forth. Conductive carbon materials with high surface areas are suitable for the rational design and construction of high-performance cathodes and have been widely researched for the application in Li-S cells.[4,36,37] However, to satisfy the requirements of practical application, the carbonaceous hosts for sulfur of Li-S battery cathodes should meet the following demands[9,38-40]: (a) hierarchical interconnected channels to enable fast transmission of electron and Li ions in redox reactions, (b) strong interactions with the reductive species of sulfur to avoid the harmful dissolution of the lithium polysulfide within the electrolyte, (c) high electrical conductivity to make sure the porous network in composite cathode for efficient utilization of sulfur, (d) large pore volume to ensure high sulfur loading and outstanding cycling property to accommodate the volume expansion during cycling. Frankly speaking, the intrinsic properties of various carbonaceous hosts cannot completely meet all the above demands by their own. The carbon materials based cathodes always suffer from poor cycle stability resulted from the insufficient interaction between polar hydrophilic lithium polysulfides and nonpolar hydrophobic carbonaceous materials.[41,42] Therefore, it is highly desirable to design and construct composite carbon-based nanostructures as sulfur hosts combining the advantages of individual components.[43] Fortunately, previous advances indicated that a large amount of polar metal oxides including: TiO_x,[44-46] MnO_2,[47,48] SiO_2,[49,50] Al_2O_3,[51,52] and MgO[53,54] drew increasing interests for hosting sulfur. One of the main reason is that the diffusion process of lithium polysulfides to the electrolyte can be suppressed through their strong bonds with metal oxides, resulting in sufficiently good confinement for the intermediate polysulfides.[55,56]

Recent studies have shown major advancements of carbon-metal oxide nanocomposites as cathodes in the field of rechargeable Li-S batteries. To give a comprehensive analysis of how the conductive carbon combines with metal oxides increase the sulfur utilization and improves the cycling performance of current Li-S batteries, this chapter summarized the progresses in the carbon-metal oxide composites for Li-S battery cathodes in detail. Finally, we gave the insights to clarify the requirements and future directions for further improvements and practical application of Li-S cells.

2. Graphene-Metal Oxide Nanocomposites

Graphene, a single layer of sp^2-bonded carbon atoms arranged in a hexagonal crystal lattice, was first discovered by Novoselov et al. in 2004.[57] It has been regarded as the promising carbon host for sulfur element due to its light weight, large surface area, excellent electrical conductivity, and high mechanical stability, etc. In the early researches, Zhang et al.[58] reported a modified graphene oxide with reactive functional groups, the interactions between sulfur/polysulfides and graphene oxide brought highest initial specific capacity of 1400 mAh g^{-1} and stable reversibility after over 50 cycles at 0.1 C. In addition, Dai et al.[59] obtained a composite cathode by fabricating uniform sulfur particles covered by poly(ethylene glycol) (PEG) containing grapheme nanosheets. It was found that graphene coating layers could improve electrical conductivity of the coated sulfur particles and accommodate volume deformation of sulfur particles during redox process.

Since individual graphene is not able to confine lithium polysulfides with strong interaction, the overall performances of Li-S batteries still suffer from insufficient stability during long-time cycling. In this regard, various kinds of graphene–metal oxide nanocomposites were prepared and studied to obtain high performance Li-S cells. Wang et al.[60] constructed N-doped graphene to use as a conductive carbon matrix for sulfur impregnation, which was further covered by atomic layers of TiO_2 with adjustable thickness (Fig. 2(a)). The theoretical calculation based upon density functional theory (DFT) confirmed the strong chemical binding between TiO_2 nanocrystal and polysulfide species. The graphene–TiO_2 hybrid cathode with the S loading of 59 wt.% delivered a reversible specific capacity of 1070 mAh g^{-1}, and the capacity retention of 85.6% over 500 cycles at a rate of 1 C were achieved (Fig. 2(b)). Similarly, Chu group[41] fabricated mesoporous TiO_2 nanocrystals on the reduced grapheme oxide (rGO) surface. As cathode for Li-S cells (72 wt.% sulfur), the TiO_2 nanocrystals can uniformly anchor the sulphur species inside it and

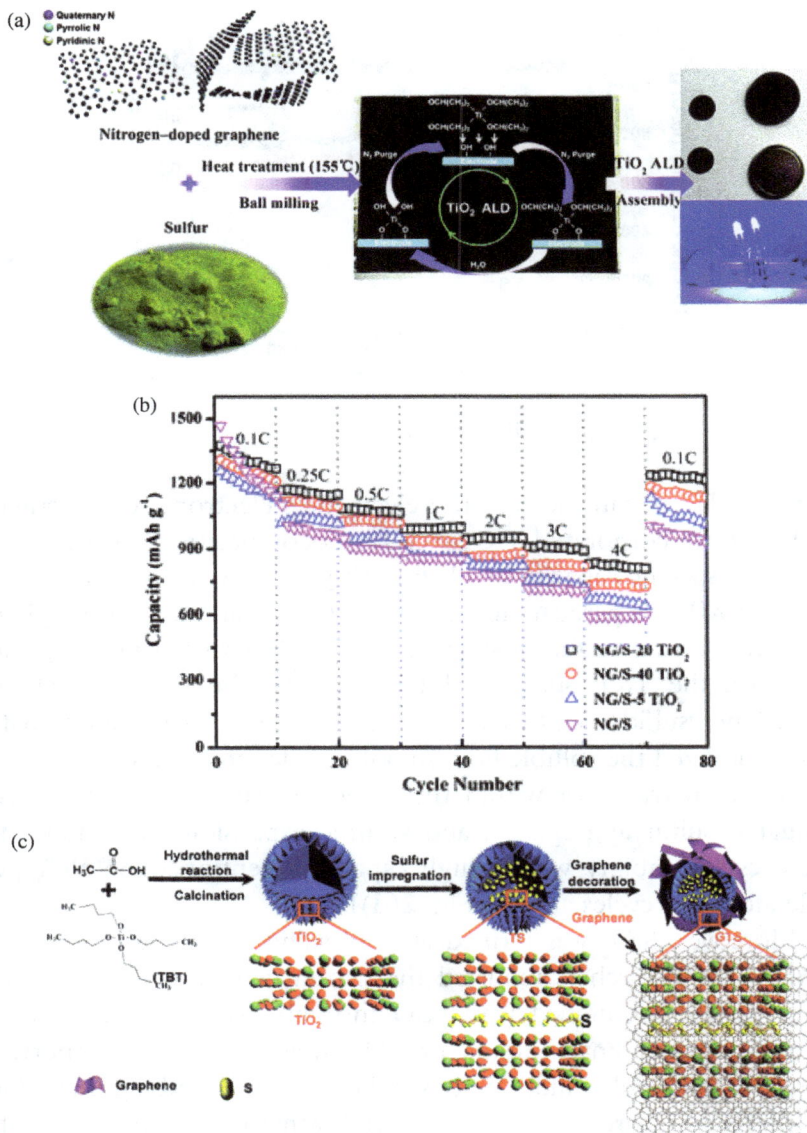

Fig. 2. (a) Preparation of the NG/S–TiO$_2$ composite. (b) Rate performances of the cathodes at various current densities.[60] (c) Schematic of the fabrication process of the graphene-modified TiO$_2$-S composite. (d) Cycling performance and corresponding coulombic efficiency for different cathodes.[61] Reprinted with permission.

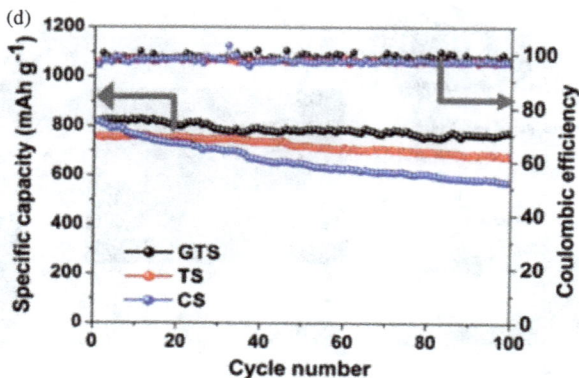

Fig. 2.   (*Continued*)

graphene layers can increase the electrode's electronic conductivity. Such sulfur-incorporated TiO_2@rGO nanocomposite demonstrated a reversible specific capacity of 1116 mAh g^{-1} at 0.2 C over 100 cycles. Wang et al.[61] prepared graphene-modified hierarchical TiO_2 sphere frameworks as a sulfur host (Fig. 2(c)). With this nanocomposite structure, the TiO_2 sphere exhibited strong chemisorption ability toward polysulfide, and rGO devoted high electrical conductivity, which localized the soluble lithium polysulfides and facilitated Li ions and electron transport within the cathodes. Such synergistic effect mitigated sulfur aggregation and volume expansion and leading to stable cycling ability with a high capacity retention of 99.96% per cycle after 400 cycles at 1 C (Fig. 2(d)).

Moreover, SiO_2 was termed as "polysulfide reservoir" by Ozkan et al.[56] The researchers reported the nanostructures of SiO_2-coated sulfur particles combined with rGO through two-step wet chemical strategy, wherein small amount of SiO_2 additive played an important role in trapping polysulfides species and effectively desorbing them during charge/discharge process. Sun et al.[62] achieved the *in situ* growth of MnO_2 nanosheets on conductive graphene surface. Due to the improved electric conductivity of MnO_2 nansheets and resolved insulating sulfur, the initial discharge capacity reached 1395 mAh g^{-1} at a current density of 0.5 C with low capacity decay of 0.3% per cycle from 10 to 100 cycles. Wang et al.[63] synthesized a porous Co_3O_4@N-C/rGO nanocomposite derived from ZIF-67 through a pyrolysis process. Such cathode with a high loading of sulfur (5.89 mg cm^{-2}) displayed stable cycling performance (611 mAh g^{-1} at a rate of 2 C over 1000 cycles). The first principle calculations further approved that the Co_3O_4@N-C/rGO nanocomposite can effectively hinder polysulfides within the cathode after multiple cycles. As promising candidate for Li-S batteries, conductive graphene with plenty of hierarchical nanopores have been studied for the past few years. Wei *et al.*[64] prepared porous graphene-$Ca(OH)_2$ by a facile chemical vapor deposition (CVD) method where precursor CaO was used as both the template and catalyst. The as-obtained cathode material showed low interfacial resistance, strong surface entrapment, robust framework, rapid mass transport, and short ion diffusion pathways. Thus delivering a discharge capacity of 357 mAh g$^{-1}_{cathode}$ (656 mAh g$^{-1}_{sulfur}$) at a high rate of 5.0 C.

3. CNT-Metal Oxide Nanocomposites

CNTs with anisotropic one-dimensional (1D) nanostructure and high surface area, can provide abundant active sites and enhance the electrical conductivity of S cathode.[65,66] Jin *et al.*[67] employed a simple precipitation method to infiltrate sulfur into the large pore of MWCNTs. The hybrid cathode exhibited high sulfur utilization but poor cycling performance after 30 cycles with just 63% capacity retention. This may indicate that CNTs have a limited function to absorb the soluble polysulfide intermediates and then diffuse into the electrolyte. Incorporation of conductive CNTs network with polar metal oxides proved to be a reliable solution to this problem. Lee *et al.*[68] filled the MWCNTs with ordered tin-monoxide (SnO) anoparticles. After impregnating with high loading of sulfur (70 wt.%), the conductive MWCNTs-SnO-S ternary hybrids showed small volume change during cycling and good ability to capture polysulfides, contributing to high initial specific capacity of 1682.4 mAh · g^{-1} at 0.1 C and the capacity of 530.1 mAh · g^{-1} was retained (0.5 C after 1,000 cycles). Dong *et al.*[69] manufactured a nanocomposite of CNTs modified with MnO nanoparticles as hosts for sulfur. In Li-S cells, such binary hybrid cathode presented a better cycling stability more than 100 cycles compared to the CNTs cathode. The CNTs/MnO cathode also demonstrated a higher specific capacity of 716 mAh g^{-1} compared to CNTs cathode (415 mAh g^{-1}) at a current density of 5.0 C. Such enhanced electrochemical performance can be ascribed to the synergistic effects of the CNTs and MnO. CNTs act as good conductor and physical barrier to sulfur and MnO helps to bind sulfur via Mn–S or S–O bonds to prevent the dissolution of polysulfides.

The next generation of miniaturized and flexible electronics devices are eager for energy storages that are deformable and shape conformable.[70,71] One of the critical challenges for designing flexible Li-S cells are development of high capacity cathode with robust conductive network and stable cycling performance.[72,73] Inspired by this, a three dimensional (3D) flexible cathode was fabricated comprised of N-doped carbon foam@CNTs covered with uniform MgO nanoparticles (Fig. 3(a)).[74] This conductive CNTs layers can strengthen the flexibility and provide porous framework for efficient electronic/ionic transport. Furthermore, the N-doping and MgO nanoparticles can significantly enhance chemisorption of polysulfide species and suppress the shuttling effect. As the hosts for sulfur, the hybrid cathodes manifested a reversible areal capacity of 10.4 mAh cm^{-2}, and retained 8.8 mAh cm^{-2} over 50 cycles at a current density of 0.05 C. Zhang et al.[75] constructed a multifunctional CNTs paper/TiO$_2$ barrier which effectively suppressed the diffusion of lithium polysulfides to the electrolyte, thus enhancing the cycling ability of Li-S cells. The electrode with 70% sulfur loading exhibits excellent cycling ability for more than 250 cycles at a rate of 0.5 C. Wang et al.[76] developed sandwich-like MnO$_2$/GO/CNTs interlayers as polysulfide-trapping shields for Li-S cell cathodes (Fig. 3(b)). The superaligned CNT films combined with GO enhance the overall electrical conductivity, and MnO$_2$ nanoparticles provided effective chemical adsorption of polysulfides. The synergetic effect of the MnO$_2$/GO/CNTs nanocomposite enabled the constructed Li-S batteries with the high sulfur content (60–80 wt.%) slow capacity decay (0.029% per cycle after 2500 cycles at 1 C) (Fig. 3(c)).

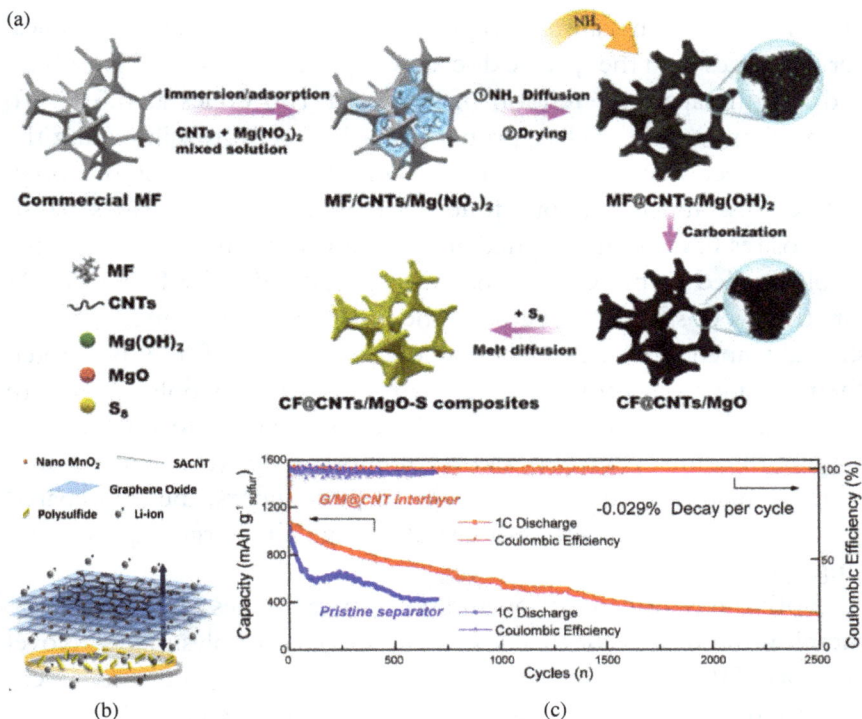

Fig. 3. (a) Schematic of the preparation process for CF@CNTs/MgO-S composite.[74] (b) Schematic diagram of the functional CNTs paper/TiO$_2$ composite. (c) Prolonged cycling performance of electrodes at a rate of 1 C.[76] Reprinted with permission.

4. Hollow Carbon Nanosphere (HCNPs)-Metal Oxide Nanocomposites

As cathodes for Li-S cells, HCNPs can not only improve the overall electronic conductivity of the nanocomposites but also have sufficient pore volume for sulfur impregnation, which helps to realize ideal reversible capacity due to high utilization of active materials.[33,77] Nazar *et al.*[78] precisely tailored both porosity and robust shell properties of HCNPs, and provided abroad view of the opportunities and limitations of HCNPs as cathodes for Li-S batteries. Archer group[79] reported mesoporous HCNPs that trap active materials in interior and porous shell. As the electrode for rechargeable Li-S cells, the

HCNPs@S carbon capsules represented promising electrochemical performances with the specific discharge capacity of up to 1071 mAh g^{-1} and a high capacity retention of 91% over 100 cycles at 0.5 C. To further improve the cycling properties of Li-S cells by minimizing the shuttle effect that physically and chemically dual-prevent the polysulfides dissolution, various functional HCNPs-metal oxides nanocomposites have been designed and investigated. He et al.[80] reported a yolk-shelled carbon@Fe$_3$O$_4$ nanobox as cathodes for Li-S cells. As for this yolk-shelled structure, the polysulfide intermediates are strongly anchored inside the nanobox because of the promoted chemical binding between Fe$_3$O$_4$ core and lithium polysulfides. In addition, the sufficient space within carbon shell can enable high loading of sulfur (80 wt.%) and accommodate the volume expansion of active materials during cycling. With these merits, the as-prepared nanocomposite cathodes demonstrated high discharge capacity and long cycle life.

Both the N-doped carbon and polar metal oxides can enhance the chemical interaction of the host with sulfur and polysulfides. Novel N-doped HCNPs coated with MnO$_2$ nanosheets architectures were constructed as hosts for sulfur.[81] Thanks to the improved electron conductivity of N-doped HCNPs, and the strengthened electron/lithium-ion transport from MnO$_2$ nanosheet, the HCNPs@MnO$_2$/S cathode with the sulfur loading of 70 wt.% demonstrated a capacity retention of 93% over 100 cycles at 5 C (Figs. 4(a) and 4(b)). Similarly, Chen et al.[82] put forward a rational design of N-doped HCNPs covered with MnO$_2$ nanosheets on both sides (Fig. 4(c)). The physical and chemical dual-encapsulation solution to this HCNPs@MnO$_2$ nanocomposite allowed for high performance Li-S cells. It exhibited a high initial capacity of 1249 mAh g^{-1} and stable cycle life with a small capacity decay of 0.041% per cycle at 0.5 C over 1000 cycles (Fig. 4(d)).

The multifunctional and integrated HCNPs@metal oxides cathodes can embody high content of active material, provide sufficient electron-modified interfaces, and effectively restrain the dissolution of

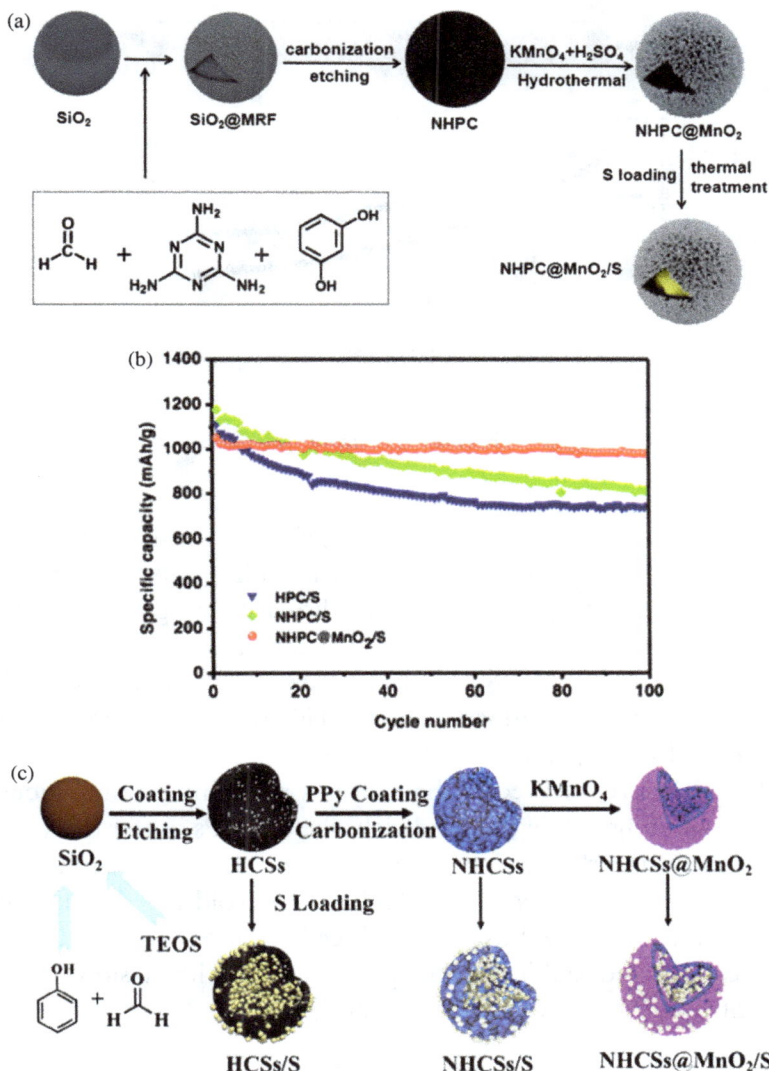

Fig. 4. (a) Schematic illustration showing the synthesis of NHPC@MnO$_2$/S nanocomposite. (b) Discharge capacities for various composite cathodes over 500 cycles at 0.5 C.[81] (c) Fabrication of HCSs/S, NHCSs/S and NHCSs@MnO$_2$/S composites. (d) Cycling ability and coulombic efficiency of three cathodes at 0.2 C.[82] Reprinted with permission.

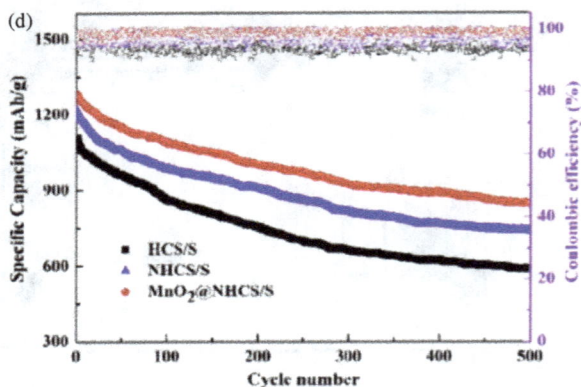

Fig. 4. (*Continued*)

polysulfide species via the synergistic effect. Lou group[83] prepared conductive polar TiO@HCNPs as the sulfur host. The TiO@HCNPs/S nanocomposite electrode displayed a specific discharge capacity of 1100 mAh g^{-1} at 0.1 C, and stable cycle ability with the capacity decay of only 0.08% per cycle after 500 cycles at 0.2 C. Fang *et al.*[84] proposed a sandwich-like C@TiO$_2$@C hollow microsphere with both physical and chemical dual-restriction for polysulfides. The interlay TiO$_2$ layers served as the carrier for confining polysulfides by chemical binding, while the sandwich-like HCNPs buffer the volume expansion during the cycling. Rehman *et al.*[85] designed a unique Si/SiO$_2$ cross-link with hierarchical HCNPs to avoid the dissolution of polysulfides. As sulfur host for Li-S cells, the Si/SiO$_2$@HCNPs/S cathode delivered stable cycling performance with a slow capacity decay of 0.063% per cycle after 500 cycles at 2 C.

5. Porous Carbon (PC)-Metal Oxide Nanocomposites

Porous Carbon (PC), with excellent electrical conductivity, high specific surface area, unable pore sizes and abundant interconnecting frameworks, has been considered as promising cathodes for Li-S cells.[79,86,87] PC can greatly improve conductivity of the active materials, accommodate volume expansion of sulfur, and encapsulate the dissolved lithium polysulfides during charge/discharge cycling.[88] Schuster *et al.*[66] fabricated spherical ordered mesoporous carbon (OMC) nanoparticles with the diameter of 300 nm, which showed high surface areas of 2445 m^2 g^{-1} with plenty of large and small mesopores. In Li-S cells, the OMC/S spherical cathodes exhibited a specific discharge capacity of 1200 mAh g^{-1}. Li *et al.*[89] reported an *in situ* strategy to prepare 3D porous graphitic carbon (PGC) with uniform sulfur nanoparticles. The sulfur loading of this 3D PGC@S nanocomposite can be tuned to an ultrahigh level (90 wt.%), leading to high reversible capacity of 1382 mAh g^{-1} at a current rate of 0.5 C.

Increasing interests have been devoted to the rational construction of PC@metal oxides hybrid nanostructures and applied to the Li-S battery cathodes. B_2O_3-coated carbon microtubes with highly porous structure were prepared through a thermal-treated process.[89] Such cathode displayed an initial capacity of 140.7 mAh g^{-1} over 5000 cycles at a high rate of 100 C. The first principle calculation revealed the strong chemical binding between lithium polysulfide species and the B_2O_3 nanocrystals, resulting in such stable cycling ability (Figs. 5(a) and 5(b)). Tao *et al.*[90] updated the confinement approach through integrating mesoporous carbon framework with Nb_2O_5 nanocrystals, which combined the merits of the physical trap and the chemical interaction with sulfur and polysulfides (Fig. 5(c)). The electrochemical kinetic measurement further confirmed that the redox reaction kinetics of polysulfide that can be greatly accelerated by the Nb_2O_5 nanocrystals, especially for the reduction process of soluble Li_2S_4/Li_2S_6 to insoluble Li_2S/Li_2S_2 (Fig. 5(d)). With this regard, the nanocomposite cathode showed a high reversible capacity of 1289 mAh g^{-1} and achieved a specific capacity of 913 mAh g^{-1} over 200 cycles at a rate of 0.5 C. From the report of An *et al.*,[91] a unique nanostructure with ultrafine TiO_2 nanocrystal confined in N-doped porous carbon (NPC) derived from an MOF precursor was constructed. Such TiO_2@NPC nanocomposite acted as efficient polysulfides barrier can enhance the kinetic of polysulfide redox reactions (Fig. 5(e)). As the cathode for Li-S cells, TiO_2@NPC cathode delivered the high specific capacity of 1460 mAh g^{-1} combined with a capacity retention of 71% over 500 cycles at a current rate of 0.2 C.

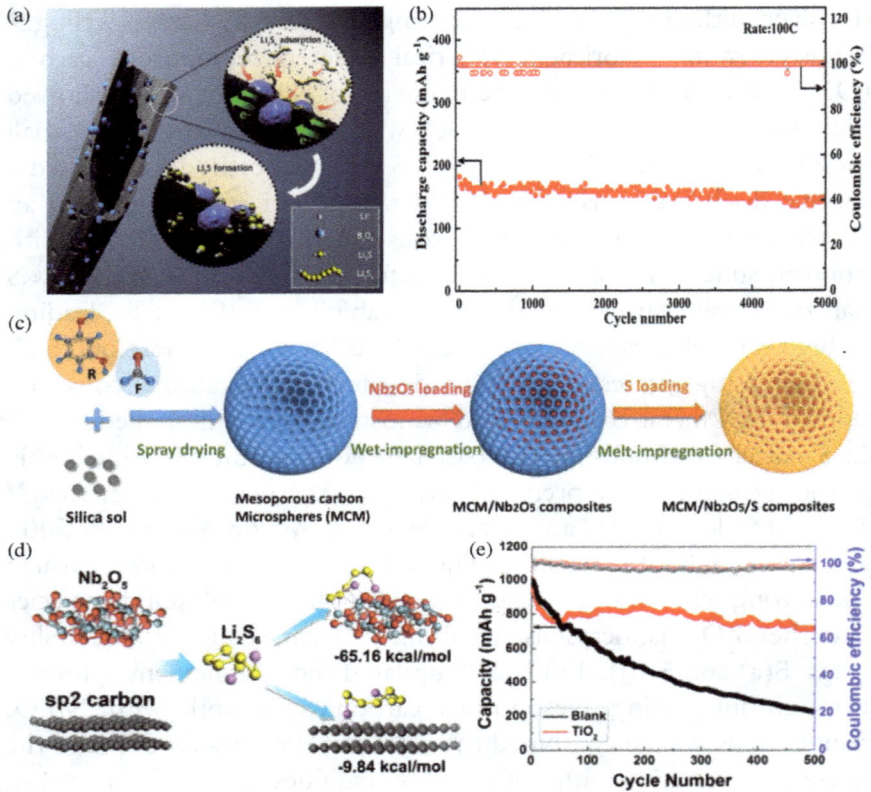

Fig. 5. (a) A schematic illustration of carbon microtubes modified with B$_2$O$_3$ nanocrystal. (b) Cycle performance of B$_2$O$_3$-carbon microtube composite cathode.[89] (c) Synthesis of the mesoporous carbon/Nb$_2$O$_5$/S composite. (d) Binding geometric configurations Li$_2$S$_6$ with Nb$_2$O$_5$ derived from *ab initio* calculations.[90] (e) Schematic of TiO$_2$@NC crystals and the charge/discharge process of the as-prepared Li-S cells.[91] Reprinted with permission.

6. Conclusions and Perspective

In summary, we have attempted to review the recent advances in carbon-metal oxide-based cathodes of Li-S cells. With improved electronic conductivity, rapid ion/electron transport and effective trap of polysulfide species, various carbon-metal oxide nanocomposite cathodes have been intensively investigated. Carbon-metal oxides have proved to be ideal candidates for high-performance Li-S batteries in terms of reversible capacity, cycle ability and rate capability.

In the next research, to obtain Li-S batteries with high energy density and stable cycling stability toward commercial applications, the design and development of carbon-metal oxide/S based cathodes need to meet the following demands: (1) high sulfur content, a large amount of macropores and mesopores are necessary in the carbon-metal oxides/S nanocomposites. (2) To ensure high capacity release of S, the host carbon-metal oxides are expected to have excellent electrical conductivity. (3) To enhance the cycling ability, sufficient contact areas between the metal oxides and the lithium polysulfides are needed, enabling physical and chemical dual-restriction for dissolution of the active materials. (4) For flexible Li-S batteries, the binder-free carbon-metal oxide/S cathodes with minimizing usage of binders and additional conductive agents are desirable. Although the steps for carbon-metal oxides/S composite cathode to practical applications encounter many challenges, carbon-metal oxides are regarded as promising hosts for sulfur due to their comprehensive superiority. The recent progress in this chapter brings great prospects of near future commercialization for carbon-metal oxides/S based Li-S battery cathodes.

Acknowledgments

This project is supported by the Ministry of Science and Technology of China (2016YFA0201904) and National Natural Science Foundation of China (21631002, U1632119, and 91333105).

References

1. D. Herbert, *J. Ulam* (1962).
2. R. Rauh *et al.*, *J. Electrochem. Soc.* **126**, 523 (1979).
3. P. G. Bruce *et al.*, *Nature Mater.* **11**, 19 (2012).
4. Y. X. Yin *et al.*, *Angew. Chem. Int. Ed.* **52**, 13186 (2013).
5. D. S. Jung *et al.*, *Nano Lett.* **14**, 4418 (2014).
6. S. Evers and L. F. Nazar, *Acc. Chem. Res.* **46**, 1135 (2012).
7. A. Manthiram *et al.*, *Chem. Rev.* **114**, 11751 (2014).
8. S. Rehman *et al.*, *J. Mater. Chem. A* **5**, 3014 (2017).
9. Z. Li *et al.*, *Energy Environ. Sci.* **9**, 3061 (2016).
10. Y. Yang *et al.*, *Chem. Soc. Rev.* **42**, 3018 (2013).

11. J. Song *et al.*, *Angew. Chem. Int. Ed.* **54**, 4325 (2015).
12. G. C. Li *et al.*, *Adv. Energy Mater.* **2**, 1238 (2012).
13. Y. Mao *et al.*, *Nat. Commun.* **8**, 14628 (2017).
14. C. Huang *et al.*, *Nat. Commun.* **5**, 3015 (2014).
15. X.-B. Cheng *et al.*, *Energy Storage Mater.* **6**, 18 (2017).
16. S. Jiang *et al.*, *Chem-Asian J* **13**, 1379 (2018).
17. Z. Lin *et al.*, *Angew. Chem.* **125**, 7608 (2013).
18. B. D. Adams *et al.*, *Nano Energy* **40**, 607 (2017).
19. K. K. Fu *et al.*, *Energy Environ. Sci.* **10**, 1568 (2017).
20. Z. Wang *et al.*, *Nat. Commun.* **5**, 5002 (2014).
21. G. Zhou *et al.*, *Nano Energy* **11**, 356 (2015).
22. G. Hu *et al.*, *Adv. Mater.* **28**, 1603 (2016).
23. Z. Du *et al.*, *ACS Appl. Mater. Interfaces* **9**, 43696 (2017).
24. G. Xu *et al.*, *Energy Environ. Sci.* **10**, 2544 (2017).
25. Y.-S. Su and A. Manthiram, *Chem. Commun.* **48**, 8817 (2012).
26. Z. Guo *et al.*, *Electrochim. Acta* **230**, 181 (2017).
27. Z. Li *et al.*, *ACS Nano* **8**, 9295 (2014).
28. S. Liu *et al.*, *Adv. Mater.* **30**, 1706895 (2018).
29. H. Yu *et al.*, *Nano Res.* **10**, 2495 (2017).
30. J. Zang *et al.*, *Nano Res.* **8**, 2663 (2015).
31. Y. Zhao *et al.*, *Adv. Mater.* **26**, 5113 (2014).
32. Y. Zhong *et al.*, *J. Mater. Chem. A* **4**, 9526 (2016).
33. N. Brun *et al.*, *Phys. Chem. Chem. Phys.* **15**, 6080 (2013).
34. W. Zhou *et al.*, *Nano Lett.* **14**, 5250 (2014).
35. J. Zhang *et al.*, *Angew. Chem. Int. Ed.* **55**, 3982 (2016).
36. L. Ma *et al.*, *ACS Nano* **10**, 1050 (2015).
37. D.-W. Wang *et al.*, *J. Mater. Chem. A* **1**, 9382 (2013).
38. A. Manthiram *et al.*, *Adv. Mater.* **27**, 1980 (2015).
39. J. Zhang *et al.*, *Small Methods* **2**, 1700279 (2018).
40. R. Fang *et al.*, *Adv. Mater.* **29**, 1606823 (2017).
41. Y. Li *et al.*, *ACS Appl. Mater. Interfaces* **8**, 23784 (2016).
42. H. J. Peng and Q. Zhang, Angew. *Chem. Int. Ed.* **54**, 11018 (2015).
43. Z. Li *et al.*, *Carbon* **92**, 41 (2015).
44. B. Ding *et al.*, *Electrochim. Acta* **107**, 78 (2013).
45. X. Tao *et al.*, *Nano Lett.* **14**, 5288 (2014).
46. X. Z. Ma *et al.*, *J. Electroanal. Chem.* **736**, 127 (2015).
47. Z. Li *et al.*, *Angew. Chem. Int. Ed.* **54**, 12886 (2015).
48. J. Zhang *et al.*, *Nano Lett.* **16**, 7276 (2016).
49. P. Wei *et al.*, *Int. J. Hydrogen Energy* **41**, 1819 (2016).

50. S.-K. Lee *et al.*, *Nano Lett.* **15**, 2863 (2015).
51. Z. Zhang *et al.*, *Electrochim. Acta* **129**, 55 (2014).
52. X. Han *et al.*, *Nano Energy* **2**, 1197 (2013).
53. X. Tao *et al.*, *Nat. Commun.* **7**, 11203 (2016).
54. R. Ponraj *et al.*, *ACS Appl. Mater. Interfaces* **8**, 4000 (2016).
55. A. Eftekhari, D.-W. Kim, *J. Mater. Chem.* A **5**, 17734 (2017).
56. B. Campbell *et al.*, *Nanoscale* **7**, 7051 (2015).
57. K. S. Novoselov *et al.*, *Science* **306**, 666 (2004).
58. L. Ji *et al.*, *J. Am. Chem. Soc.* **133**, 18522 (2011).
59. H. Wang *et al.*, *Nano Lett.* **11**, 2644 (2011).
60. M. Yu *et al.*, *Energy Environ.* Sci. **9**, 1495 (2016).
61. L. Gao *et al.*, *J. Mater. Chem.* A **4**, 16454 (2016).
62. W. Sun *et al.*, *Electrochim. Acta* **207**, 198 (2016).
63. J. Xu *et al.*, *J. Mater. Chem.* A **6**, 2797 (2018).
64. C. Tang *et al.*, *Adv. Funct. Mater.* **26**, 577 (2016).
65. J. Guo *et al.*, *Nano Lett.* **11**, 4288 (2011).
66. J. Schuster *et al.*, *Angew. Chem. Int. Ed.* **124**, 3651 (2012).
67. W. Ahn *et al.*, *J. Power Sources* **202**, 394 (2012).
68. A.-Y. Kim *et al.*, *Nano Res.* **10**, 2083 (2017).
69. T. An *et al.*, *J. Mater. Chem.* A **4**, 12858 (2016).
70. X. Peng *et al.*, *Chem. Soc. Rev.* **43**, 3303 (2014).
71. Y. H. Kwon *et al.*, *Adv. Mater.* **24**, 5192 (2012).
72. Z. Yuan *et al.*, *Adv. Funct. Mater.* **24**, 6105 (2014).
73. K. Jin *et al.*, *J. Phys. Chem.* C **117**, 21112 (2013).
74. M. Xiang *et al.*, *Adv. Funct. Mater.* **27**, (2017).
75. G. Xu *et al.*, *Nano Res.* **8**, 3066 (2015).
76. W. Kong *et al.*, *Adv. Funct. Mater.* **27**, (2017).
77. S. Chen *et al.*, *J. Mater. Chem.* A **2**, 16199 (2014).
78. G. He *et al.*, *ACS Nano* **7**, 10920 (2013).
79. N. Jayaprakash *et al.*, *Angew. Chem. Int. Ed.* **123**, 6026 (2011).
80. J. He *et al.*, *Adv. Mater.* **29**, 1702707 (2017).
81. X. Zhang *et al.*, *Nanotechnol.* **28**, 475401 (2017).
82. M. Chen *et al.*, *Chem. Eng. J.* **335**, 831 (2018).
83. Z. Li *et al.*, *Nat. Commun.* **7**, 13065 (2016).
84. M. Fang *et al.*, *J. Mater. Chem.* A **6**, 1630 (2018).
85. S. Rehman *et al.*, *Adv. Mater.* **28**, 3167 (2016).
86. X. Ji *et al.*, *Nat. Mater.* **8**, 500 (2009).
87. W. Hu *et al.*, *Funct. Mater. Lett.* **9**, 1650015 (2016).
88. G. Li *et al.*, *Nat. Commun.* **7**, 10601 (2016).

89. Z. Su *et al.*, *J. Mater. Chem.* A **4**, 8541 (2016).
90. Y. Tao *et al.*, *Energy Environ. Sci.* **9**, 3230 (2016).
91. Y. An *et al.*, ACS *Appl. Mater. Interfaces* **9**, 12400 (2017).

Chapter 3

Recent Advances of Polar Transition-Metal Sulfides Host Materials for Advanced Lithium–Sulfur Batteries

Liping Chen*, Xifei Li*,§, and Yunhua Xu[†,‡,¶]

Lithium sulfur batteries (LSBs) have been one of the most promising second batteries for energy storage. However, the commercialization of LSBs is still hindered by low sulfur utilization and poor cycling stability, resulting from shuttle effect and low redox kinetics of lithium polysulfides (LiPSs). Significant progress has been made over the years in enhancing the batteries performances and tap density with the transition-metal sulfides as sulfur host or additive in LSBs. In this review, we present the recent advances in the use of various nanostructured transition-metal sulfides applied in LSBs, and also focus on the interaction mechanisms of polar transition-metal sulfides with LiPSs and its catalysis for the redox of LiPSs. It may provide avenues for the application of transition-metal sulfides in LSBs. The challenges and perspectives of transition-metal sulfides are also addressed.

*Institute of Advanced Electrochemical Energy and School of Materials Science and Engineering, Xi'an University of Technology, Xi'an 710048, P. R. China.
†Collaborative Innovation Center of Modern Equipment and Green Manufacturing, Xi'an University of Technology, Xi'an 710048, P. R. China.
‡Yulin University, 4 Chongwen Road, Yulin, Shaanxi 719000, P. R. China.
§xfli2011@hotmail.com
¶xuyunhua@vip.163.com

47

Keywords: Lithium sulfur batteries; transition-metal sulfides; shuttle effect; redox kinetic.

1. Introduction

Lithium sulfur batteries (LSBs) hold great prospect to meet the increasing need of advanced energy storage in battery field, owing to their high theoretical specific capacity (1675 mAh g^{-1}), low cost, and environmental friendliness.[1-3] However, the application of LSBs has been challenged by several obstacles, particularly low active material utilization and poor cycling stability. It may be due to the poor electronic conductivity of sulfur and its discharge end product Li$_2$S, the shuttle effect of the dissolved intermediates lithium polysulfides (LiPSs), and the large and repeated volume change.[4] Among these issues, the shuttle effect results in low utilization of sulfur, low Coulombic efficiency, and capacity decay, as well as anode corrosion,[5] which is the most thorny problem that leads to capacity decay.[6,7] Tremendous strategies have been proposed to retard the shuttle effect through using carbon materials as the sulfur host,[8,9] modification of the separator,[10,11] even designing interlayer or cathode coating.[12,13] However, non-polar carbon materials only show limited anchoring sites for LiPSs and weak interaction between them, eventually leading to poor battery performance.[5,14] In recent years, the concept of using polar conductive materials for immobilizing sulfur and LiPSs by strong chemical bonding is an acknowledged strategy to suppress the diffusion of LiPSs and enhance the sulfur utilization.[7,15] In this concern, functionalized carbon materials, functional conductive polymers, and transition metal compounds, including oxide, sulfides, carbides, nitride, hydroxides and metal organic frameworks with abundant adsorption sites are regarded as promising sulfur hosts to chemically adsorb LiPSs for improving cathode performance.[5,15-17] Moreover, several studies have shown that accelerating the conversion kinetics of the LiPSs to Li$_2$S$_2$/Li$_2$S, and vice versa, is a promising way to mitigate the LiPSs shuttling.[18,19] In this concern, transition-metal sulfides (MSs, M = Mo, Co, Ti, Ni, Fe, Cu, Mn, V, W and Zr) show great promise as a kind of sulfur host benefitting from their strong

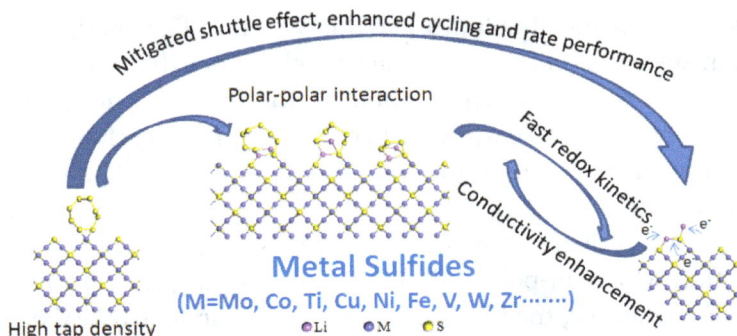

Fig. 1. Schematic illustration of the positive effect of transition-metal sulfides as sulfur host applied in LSBs.

chemical interaction with sulfur species, triggering extensive attention and researches for LSBs. As demonstrated in Fig. 1, transition-metal sulfides not only possess strong chemical affinity with LiPSs due to their large polar surface, but also reveal high catalytic activity to promote the redox kinetics of sulfur species, potentially suppressing the LiPSs shuttle.[18] Additionally, transition-metal sulfides themselves also possess high conductivity, which can enhance the conductivity of sulfur cathode when kept in contact with sulfur. Last but not least, the high density of transition-metal sulfides endows sulfur cathode high tap density, which benefits the volume energy density of LSBs.[17,20]

Here, recent advances of transition-metal sulfides applied in LSBs are reviewed, including polar–polar interaction of transition-metal sulfides with LiPSs/Li$_2$S, conductivity enhancement, catalytic effect on the redox of sulfur species, sulfur-equivalent cathode or co-cathode material and tap density improvement, and its mechanisms are systematically discussed.

2. Positive Role of Sulfides in LSBs

2.1. *Polar–polar interaction of transition-metal sulfides with LiPSs/Li$_2$S*

Transition-metal sulfides are usually utilized as host materials as well as interlayer for polysulfide adsorption in LSBs due to the strong bonds between metal sulfides and LiPSs/Li$_2$S. The binding energy between sulfides and LiPSs/Li$_2$S are much higher than that of graphite or graphene. As shown in Table 1, the binding energy of Li$_2$S$_2$ and graphite is 0.28 eV, while that of TiS$_2$ and Li$_2$S reaches 2.99 eV.[15] More importantly, the value of Li$_2$S$_2$ and Co$_9$S$_8$ is even up to 6.06 eV, which is the highest binding energy reported for lithium sulfide.[21] Additionally, the metal sulfides can also provide chemical interaction with non-polar sulfur, thus suppressing the dissolution of sulfur. For example, the binding energy can be up to 1.09 eV between

Table 1.   Binding energies of LiPSs and different metal sulfides.

MSs	Crystal face	LiPSs molecules	Binding energies (eV)	Ref.
CoS$_2$	(111)	Li$_2$S$_4$	1.97	19
Co$_3$S$_4$	(111)	Li$_2$S$_4$	2.26	24
		Li$_2$S$_6$	1.61	
		Li$_2$S$_8$	1.68	
	(220)	Li$_2$S$_4$	2.76	
		Li$_2$S$_6$	2.18	
		Li$_2$S$_8$	2.18	
Co$_9$S$_8$	(002)	Li$_2$S$_2$	2.26	21
	(202)		3.29	
	(002)		6.06	
	(008)	Li$_2$S$_4$	2.74	
	(202)		3.41	
	(002)	Li$_2$S$_4$	3.96	
			1.71	
	(202)		1.56	
	(008)		4.10	
TiS$_2$	—	Li$_2$S	2.99	37
ZrS$_2$	—		2.70	
VS$_2$	—		2.94	

Table 1. (*Continued*)

MSs	Crystal face	LiPSs molecules	Binding energies (eV)	Ref.
Ni_2S_2	(111)	Li_2S_4	2.06	34
$(Ni_2S_2)_{0.6}(FeS)_{0.4}$	(001)	Li_2S_2	1.92	42
Ni_3S_2	(110)	S_8	1.09	22
		Li_2S_8	1.92	
		Li_2S_6	2.15	
		Li_2S_4	2.29	
		Li_2S_2	3.90	
		Li_2S	4.89	
Ni_3S_2	—	Li_2S_4	0.72	28
SnS_2	—		0.80	
FeS	—		0.87	
CoS_2	—		1.01	
VS_2	—		1.04	
TiS_2	—		1.02	
graphite	—	Li_2S_4	0.28	21
graphite	—	Li_2S_4	0.34	19

Ni_3S_2 and S_8,[22] which is much higher than that of graphene (0.30 eV) and graphene with hydroxyl (0.47 eV).[23] The effect of the interaction can be visually observed by adsorption tests, that is, LiPSs solution gradually becomes lighter or colorless after the sulfides immersed.[24,25] For example, NiS exhibits strong adsorption capability for LiPSs, as shown in Fig. 2(a). The slight shift of the Raman spectra peaks as well as the XPS speaks further confirms the electrons transfer from S_x^{2-} to Ni^{2+}, and the strong chemical interaction of Ni^{2+} in NiS with S_x^{2-} in Li_2S_x.[26] Additionally, the sulfides possess coupled interaction with Li_2S_n through $S_n^{2-} - M^{\delta+}$ and $Li^+-S^{\delta-}$ (of metal sulfides) binding.[19,21,26] The high affinity between sulfides and LiPSs mitigates its dissolution and improves the cycling stability of the sulfur cathodes. For example, the cycling performance of LSBs was visibly improved with CoS_2

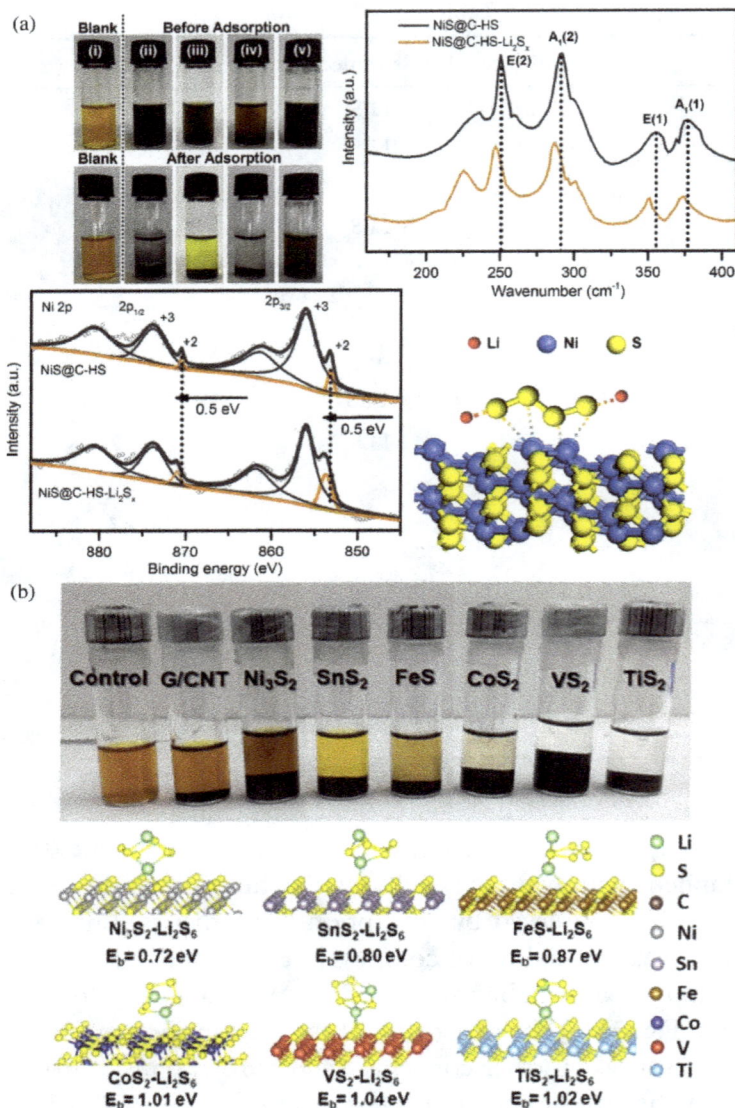

Fig. 2. (a) Photograph of (i) pristine LiPSs solution, the LiPSs solution before and after adsorbed by (ii) NiS@C–HS, (iii) C–HS@NiS, (iv) NiS–HS, and (v) C–HS; Raman and XPS spectra of Ni 2p of NiS@C–HS and NiS@C–HS–Li$_2$S$_x$; and schematic illustration of the Li$_2$S$_4$ binding configuration with NiS.[26] (b) Digital image of the adsorbed Li$_2$S$_6$ by carbon and Ni$_3$S$_2$, SnS$_2$, FeS, CoS$_2$, VS$_2$ and TiS$_2$, and simulation of Li$_2$S$_6$ interacts with these sulfides: atomic conformations and binding energies.[28] Reprinted with permission.

anchored on carbon paper as the interlayer, due to the trap ability of CoS_2 for the LiPSs by forming strong chemical interaction.[27] Compared with original carbon aerogel/S electrode, the initial specific capacity of sulfur cathode increased from 1156mAh g^{-1} to 1343mAh g^{-1} when MoS_2 (8% molar ratio) was introduced, and better stability was obtained, due to the immobilizing effect of sulfur by MoS_2.[29] In addition, Co_3S_4,[24] MoS_2,[29] Co_9S_8,[30] TiS_2,[28] SnS_2,[31] Cu_2S/CuS,[32] FeS_2,[33] NiS_2,[34] WS_2,[35] VS_2,[36] ZrS_2,[37] etc., are all capable of absorbing LiPSs. The chemical interaction is dominated by the bond of the Liþ of LiPSs and the sulfur ion of the sulfide, and the larger binding energy, the stronger interactions, thus the better anchoring effect,[28] as shown in Fig. 2(b). However, binding free energy (ΔG_B) is much more valuable to evaluate the interaction strength between cathode material and sulfur species because it possesses the advantage of involving thermal and entropy corrections.[38] Moreover, the adsorption energy (Ea) between metal sulfides and Li_2S_x/S_8 was also conducted to evaluate the chemisorption capability.[39]

It is noted that the strong interaction also facilities the formation of Li_2S. Compared to bare graphene, CoS_2 favor the controllable nucleation and growth of Li_2S.[19] Furthermore, relieved shuttle accordingly remits the corrosion of Li anode. MoS_2/graphene interlayer was designed to restrain the shuttle effect with the synergistic effect of physical and chemical adsorption. As a result, the cycled Li anode exhibited uniform and lower roughness of surface with the addition of MoS_2/graphene interlayer, due to the relieved dissolution of LiPSs to the electrolyte.[40] The strong interaction relieves the shuttle of LiPSs, which promotes the cycling stability. However, the chemical interaction is not the only critical factor for trapping LiPSs. Therefore, only the binding energy or/and binding free energy (adsorption energy) may not be the sole evaluation criteria to select the best one among these metal sulfides for performance improvement.

In addition to types of sulfides, crystal structure may affect the capacity of trapping LiPSs. This is because different crystal planes possess various adsorption energies[39] or binding energies[24] with different LiPSs molecules (Li_2S_2, Li_2S_4, Li_2S_6 and Li_2S_8). The binding energies between LiPSs and different sulfides with different crystal planes are summarized in Table 1. Consequently, sulfides should be designed with controlled crystal structure, or two different kinds of sulfides with synergistic effect for different LiPSs molecules should be composited, to maximize the interaction on different LiPSs molecules. It is noted that different crystal planes possess various binding energies for Co_9S_8. Additionally, the binding energy between Li_2S and sulfides is sitedependent, which is different in terrace and edge sites. The Mo-edge and S-edge of MoS_2 exhibit binding energy with Li_2S up to 4.48 and 2.70 eV, respectively, which are much higher than that of terrace sites (0.87 eV) of MoS_2.[41] The edges facilitate uniform distribution of Li_2S and prevent its accumulation, as a result, the sample with most edge sites delivers higher discharge capacity, cycling stability as well as rate performance than the one with terrace sites. This guides the structure design to improve cathode performance by exposing more edge sites on the surface of materials for LSBs.

2.2. Conductivity enhancement

The conductivity of sulfur host plays a pivotal role in sulphur cathode performance, which not only influences electron transport, but also tightly relates to redox kinetics of the Li–S system. If the sulfur host materials are electrical insulators, the adsorbed LiPSs may not be reduced despite off bound to the host.[43] Zhang[44] studied the relationship between the electrical conductivity of polar host materials and the redox kinetics of electrochemical reactions. The results show that conductive polar TiC increases the intrinsic activity for liquid–liquid LiPSs interconversion and liquid–solid precipitation of Li_2S more than non-polar carbon and semiconductor TiO_2.[44] Therefore, the host material must simultaneously possess interaction with LiPSs and conductivity, benefiting the homogeneous distribution of sulfur compounds, and synergistically inhibiting the LiPSs shuttle and accelerating the reaction kinetics.[43,45] In comparison to metal oxides, the transition-metal sulfides usually exhibit higher conductivity, especially for CoS2, its electrical conductivity is up to 6.7×10^5 S cm^{-1},[24] which is much higher than that of Co_9S_8 (290 S cm^{-1})[5], FeS (80 S cm^{-1}),[46]

CuS (870 S cm^{-1}),[47] reduced graphene oxide (170 S cm^{-1})[7] and Ti_4O_7 (3.2 ± 0.1 S cm^{-1}).[48] Therefore, the transition-metal sulfides can improve the overall conductivity of the sulphur cathodes, simultaneously, nanostructured transition-metal sulfides are effective conductive polar host materials which can adsorb LiPSs efficiently and mitigate Li_2S_2/LiS_2 detaching from cathodes.[5] However, the electrical conductivity of transition-metal sulfides varies significantly. They can be classified as semimetals ($CoS2$)19 and semiconductors (FeS_2).[33] Therefore, the selection of metal sulfides based on conductivity should be considered when designing sulfur host materials for LSBs.

2.3. *Catalytic effect on the redox of sulfur species*

Besides the strong chemical interaction with LiPSs, transition-metal sulfides provide some catalytic effect on the redox of LiPSs. Through accelerating the conversion of LiPSs, their diffusion to electrolyte is relieved and kinetic barriers are effectively overcome.[18] As a result, the battery performance is improved, including lower polarization, higher Coulombic efficiency, excellent cycling stability and rate performance. The redox peaks of Co_3S_4–S shifted towards the quasiequilibrium potentials and plateau gap of charge–discharge process were narrowed, due to faster redox kinetic than the pristine sulfur cathode.[24] To further testify the enhanced reaction kinetics, a symmetric cell (Fig. 3(a)) was constructed with Co_3S_4 electrodes. As demonstrated in Fig. 3(b), the current response of Co_3S_4 and acetylene black (AB) cells both become much higher after the addition of Li_2S_6. And the Co_3S_4 cell possesses much smaller semicircle reaction than the AB cell, as shown in Fig. 3(c), indicating the electrode reaction kinetics improves. Meanwhile, the Co_3S_4 cell reveals higher current responses than the AB cell in Li_2S_6-containing electrolyte, as shown in Fig. 3(d). It can be concluded that Co_3S_4 can adsorb sulfur species, simultaneously enhancing the redox reaction kinetics of LiPSs.[24] A similar result was obtained when sulfides were added into sulfur cathode, such as CoS_2,[19] WS_2.[35] A series of transition-metal sulfides were studied by Cui to investigate the factors determining the sulfur conversion chemistry.[28] The S@G/CNT cathode with the addition of VS_2, CoS_2 and

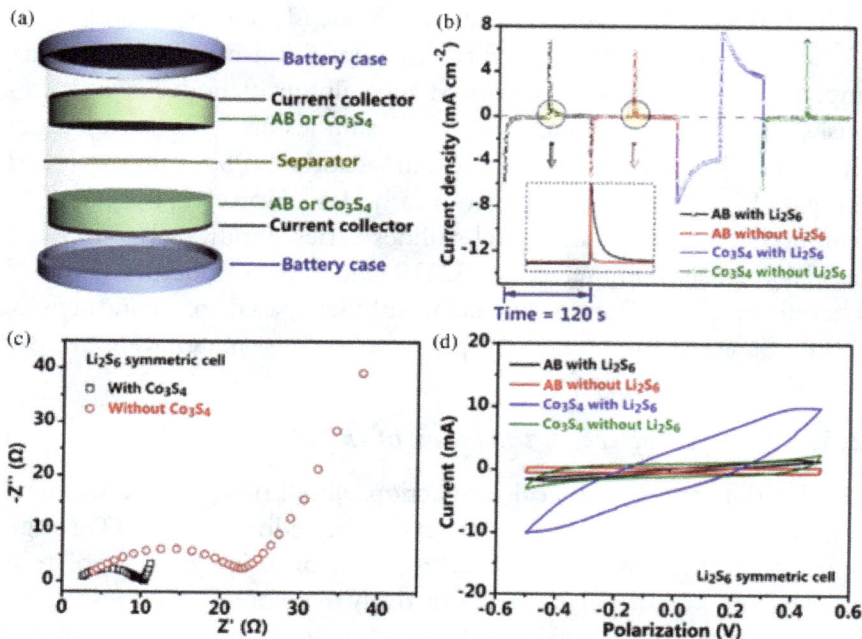

Fig. 3. (a) Schematic of a symmetric cell. (b) Chronoamperometric curves, (c) EIS plots, and (d) CV profiles of the Li2S6 and AB symmetric cells.[24] Reprinted with permission.

TiS_2 demonstrate faster Li^+ diffusion and better reaction kinetics than those of Ni_3S_2, SnS_2 and FeS (Figs. 4(a)–4(c)), indicating high catalytic activity for sulfur conversion reaction. The diffusion barriers for Li^+ on Ni_3S_2, TiS_2-containing cathodes show better reaction kinetics: a lower barrier endows faster Li^+ diffusion rate on the surface of the host materials, making redox reaction between lithium and sulfur facilitated. As a result, the cathodes containing VS_2, TiS_2 and CoS_2 with higher binding energy and lower Li^+ diffusion and activation energy barriers, deliver better capacity and cycling stability (Fig. 4(e)).

Besides the catalysis for liquid-liquid transformation of LiPSs, transition-metal sulfides promote the liquid–solid conversion process, namely, the reaction of LiPSs to Li_2S and vice versa. rGO–VS_2 is abundant in active sites for the formation of Li_2S. Furthermore, it also accelerates the conversion from Li_2S to LiPSs.[36] In contrast to the cycled rGO/S cathode with a clean rGO surface and irregular Li_2S

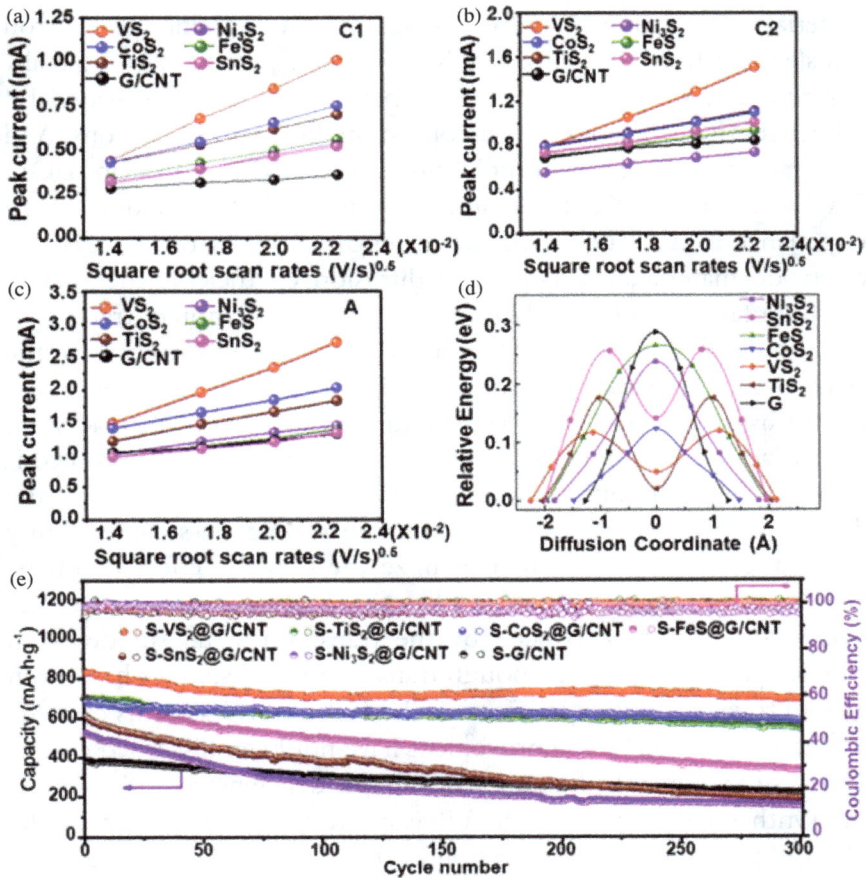

Fig. 4. Diffusion behaviors of Li$^+$ on the surface of graphene and Ni$_3$S$_2$, SnS$_2$, FeS, CoS$_2$, VS$_2$, TiS$_2$: (a) plots of first CV reduction peak current (I$_{C1}$:S$_8$ ® LiPSS), (b) plots of second CV reduction peak current (I$_{C2}$: LiPSS ® Li$_2$S$_2$/Li$_2$S), and (c) plots of CV oxidation peak current (I$_A$:Li$_2$S$_2$/Li$_2$S®S$_8$) vs. the square root of the scan rates. (d) Energy profiles of Li$^+$ diffusion behaviors on graphene and different metal sulfides. (e) Cycling stability and corresponding Coulombic efficiency at 0.5C for 300 cycles.[28] Reprinted with permission.

particles, the cycled rGO–VS$_2$/S exhibited Li$_2$S uniformly deposited on the rGO–VS$_2$ without bulk Li$_2$S particles.

The catalysis of sulfides for LiPSs conversion is closely related to their conductivity and the polar interactions with LiPSs. Upon cycling, when LiPSs maintain close electrical contact with host materials with strong chemical interaction, low resistance of electron transfer and fast kinetics of LiPSs conversion process may be really realized. The structure, stability and behavior of solvent solvated-Li$^+$ influences the behavior of the anions in the Li–S redox reactions. And the stability of the major polysulfide intermediates (S_4^{2-} in DOL: DME) can improve the reaction rates of the LSBs.[49] and Warburg impedance is associated with the diffusion of Li$^+$ in the bulk of the electrode material, the larger Li$^+$ diffusion coefficient, the smaller Warburg factor values.[50] Additionally, strong interaction of transition-metal sulfides with LiPSs leads to relatively low LiPSs viscosity in the electrolyte, inducing relatively fast Li$^+$ diffusion. Moreover, fast Li$^+$ diffusion rate facilitates the sulfur transformation chemistry on the surface of transition-metal sulfides.[28] Therefore, the electron conductivity and ion diffusion should be as high as possible in the designed cathode, and strong interaction should be simultaneously taken into consideration, to maximize the catalytic effect of LiPSs. However, the mechanisms of fast kinetics is not only related to electronic conductivity enhanced by the transition-metal sulfides with excellent conductivity.[24] Although transition-metal sulfides have been proved to exhibit catalytic effect on the LiPSs conversion, its catalytic mechanism and further improvement have not been discerned clearly. MoS$_2$ with a controlled amount of sulphur deficiency (MoS$_{2-x}$/rGO) was synthesized to catalyze the LiPSs conversion in a sulfur cathode. The CV peaks and peak separation in the MoS$_{2-x}$/rGO cell were significantlynarrowed than that of the MoS$_2$/rGO (without sulfur deficiency) and the rGO cells. Furthermore, it was demonstrated that for LiPSs transformation, more electrochemically active sites existed on the MoS$_{2-x}$/rGO surface compared to rGO, confirming sulfur deficiencies got involved in the LiPSs conversion and greatly facilitated the LiPSs redox.[51] This provides method strategy for improving the catalytic effect of sulfides.

2.4. Sulfur-equivalent cathode or co-cathode

Sulfur-equivalent cathode was proposed by Li,[52] for instance, MoS$_3$ was used as cathode instead of sulfur. The MoS$_3$ delivered an initial specific capacity of 667 mAh g^{-1} based on MoS$_3$ mass, and 1482 mAh g^{-1} based on the sulfur mass (Fig. 5(a)). It exhibited excellent cycling stability, and remained 383 mAh g^{-1} after 1000 cycles (Fig. 5(b)), due to the free of LiPSs (Fig. 5(c)). NiS was also employed as cathode with LiTFSI electrolyte for lithium batteries, and oxidation peaks and reduction peaks were observed at 1.2–1.3 V and 1.9–2.1 V, respectively, corresponding to the redox reactions of NiS. The NiS electrode exhibited the initial discharge capacities of 300 mAh g^{-1}, 450 mAh g^{-1} and 550 mAh g^{-1} at C/10, C/20 and C/50, respectively. And the fading rate was only 23% at C/6 after 100 cycles.[53] Similarly, NiS$_2$/FeS holey film cathode exhibited an excellent discharge capacity of 580 mAh cm^{-3} and a low capacity fading.[46]

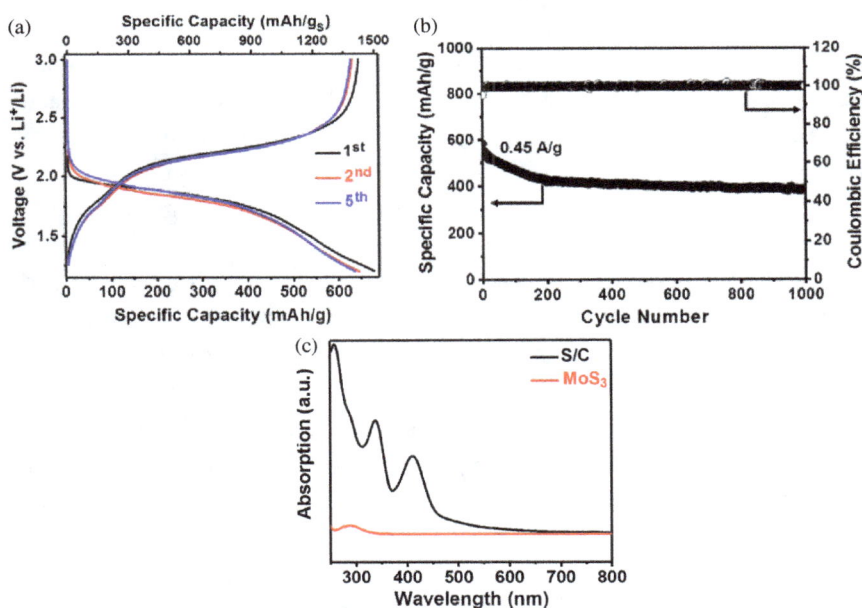

Fig. 5. Electrochemical performances of MoS$_3$ as the sulfur-equivalent cathode for LSBs. (a) Galvanostatic charge–discharge profiles at 23mAg^{-1}; specific capacity was based on the mass of sulfur in MoS$_3$ (top *x*-axis) or the total mass of MoS$_3$ (bottom *x*-axis). (b) Cycling performance and Coulombic efficiency at 0.45Ag^{-1}; (c) UV–Vis adsorption spectrum of the electrolyte of the Li–MoS$_3$ battery after 100 cycles comparing with that of conventional LSBs after 100 cycles.[52] Reprinted with permission.

TiS$_2$ was composited with carbon additives as a co-cathode material of LSBs. TiS$_2$ also suffers redox reaction with lithium, and contributed 252 mAh g^{-1} more than the pristine sulfur electrode (1082 mAh g^{-1}), because TiS$_2$ facilitates the dispersion and utilization of active material sulfur as well as stabilizes the cathode structure.[54] However, not all sulphides can be used as sulfur–equivalent cathode or co-cathode materials. For example, Co$_9$S$_8$ only exhibits a capacity of 15–35 mAh g^{-1} in the typical LSB voltage window.[21] MoS$_2$ has almost no capacity, because lithiation of MoS$_2$ occurs below 1.5V vs. Li/Li$^+$, which is lower than the voltage window of LSBs.[51,55]

2.5. *Tap density improvement*

In the face of practical application for LSBs, gravimetric energy density is also inevitable to be investigated, especially when current collector, conductive additive and binder are considered.[56] Additionally, thick electrode, long electron pathway and poor volumetric capacity are also thorny issues caused by sulfur host materials with low tap density.[57,58]

More attention has been paid to high-volume energy density for LSBs, which is a key factor when applied in compact space.[59] In recent years, great achievements have been made in promoting the cathode properties and tap density with transition-metal compounds in LSBs.[17] Transition-metal sulfides hold the great promise to obtain packed electrodes for LSBs with excellent volumetric energy density due to its high tap density, which is much higher than porous carbon.[20] The volume of rGO–VS$_2$/S samples are much smaller compared to rGO/S with the same sulfur content (Fig. 6(a)), obviously enlarging the sample density only with 5 wt.% VS$_2$. As demonstrated in Fig. 6(b), rGO–VS$_2$/S–64, rGO–VS$_2$/S–73 and rGO–VS$_2$/S–89 delivers the maximal discharge capacities of 660.0, 675.0 and 721.5mAh g^{-1}, respectively, in which the mass of conductive additive and binder are also included, and their corresponding areal discharge capacities are 1.45, 2.00 and 2.60 mAh cm^{-2}, respectively. Additionally, the volumetric discharge capacity of rGO–VS$_2$/S–89 can reach up to 1182.1mAh cm^{-3}, based on the total volume including every

(a)

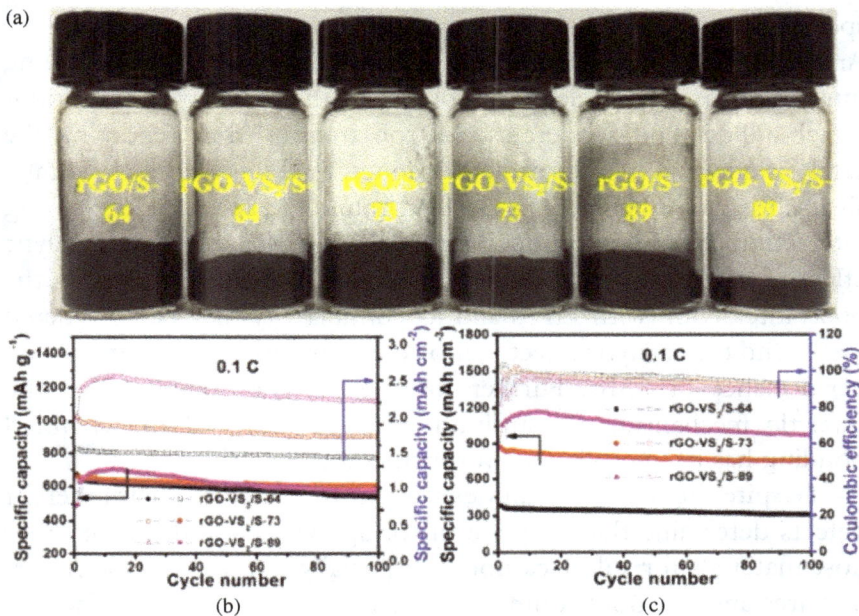

(b)

(c)

Fig. 6. (a) Volume comparison of rGO/S and rGO–VS$_2$/S with the same mass. (b) The mass/areal capacity of rGO–VS$_2$/S at 0.1 C, (c) the volumetric capacity and Coulombic efficiency of rGO–VS$_2$/S at 0.1 C.[36] Reprinted with permission.

component inside the battery (Fig. 6(c)). By contrast, rGO/S–64 only exhibits the volumetric capacity of 194.6 mAh cm^{-3}, which is far inferior to that of rGO–VS$_2$/S–64.[36] In addition to its effect on volumetric capacity, area-specific capacity has also been promoted. Extremely high areal mass loading of sulfur even reached 40 mg cm^{-2} with high areal capacity and high specific capacity using S$_8$/TiS$_2$ hybrid foam as sulfur cathode, benefitting from its high conductivity and strong affinity for LiPSs.[56]

3. Conclusions and Outlook

Combining the chemical adsorption, conductivity enhancement and catalytic effect, transition-metal sulfides can promote whole solid–liquid–solid redox kinetics of sulfur cathode, thus delivering high specific capacity, excellent cycling stability as well as rate capability. And there is a close relationship between these aspects. The strong interaction maintains electrical contact of the LiPSs with transition-metal sulfides and accelerates electron transfer, thus rendering the catalytic effect of transition-metal sulfides. Simultaneously, the catalysis itself is closely related to electron transfer. Therefore, the strong interaction and high conductivity positively promotes the catalytic effect of transition-metal sulfides for LiPSs conversion. As a result, the polar interaction with LiPSs and its conductivity may be considered firstly, and the catalytic effect is also considered for enhancing cathode performance of LSBs. Furthermore, the relationship between the catalytic mechanism of transition-metal sulfides and the chemical bonding has been challenged for further studies.

Despite the merits of sulfides as mentioned above, their inherent defects determine that they are fit for applying as additive of other host materials for sulfur cathodes, especially due to their poor surface area and low pore volume. In addition, the obtained sulfides are usually large particles, thus only limited active sites of affinity and catalysis for LiPSs. Consequently, sulfides are often composited with other sulfur host materials, such as carbon nanotube and graphene, to improve surface area, conductivity and physical confinement. Additionally, structural design is quite essential for achieving synergistic effect to improve cathode performance of LSBs. As for the structural design, nanocrystallization and quantization is an effective approach to enlarge surface and increase the active sites. Sulfides-embedded carbon polyhedron composites, which are synthesized using metal organic framework materials as precursor and template, may be ideal composites for sulfur host. The reserved structure with enough void space not only ensures the sulfur loading and accommodate the volume changes during the cycles, but also ensures uniform dispersion of transition-metal sulfides. All in all, to rationally design and utilize transition-metal sulfides may suppress the LiPSs shuttling and accelerate the reaction of sulfur species in LSBs, but it has been challenging. Some future works are required to address the catalyze mechanism of transition-metal sulfides for LSBs.

References

1. X. L. Ji, K. T. Lee and L. F. Nazar, *Nat. Mater.* **8**, 500 (2009).
2. Y. X. Yin *et al.*, Angew. *Chem. Int. Ed.* **52**, 13186 (2013).
3. A. Manthiram *et al.*, *Chem. Rev.* **114**, 11751 (2014).
4. J. Q. Huang *et al.*, *ACS Nano* **9**, 3002 (2015).
5. X. Liu *et al.*, *Adv. Mater.* **29**, 1601759 (2017).
6. Q. Pang *et al.*, *J. Electrochem. Soc.* **162**, A2567 (2015).
7. Z. Y. Wang *et al.*, *Nat. Commun.* **5**, 5002 (2014).
8. D. W. Wang *et al.*, *J. Mater. Chem.* A **1**, 9382 (2013).
9. M. P. Yu *et al.*, *Energy Storage Mater.* **1**, 51 (2015).
10. P. Han and A. Manthiram, *J. Power Sources* **369**, 87 (2017).
11. Y. B. He *et al.*, *Dalton Trans.* **47**, 6881 (2018).
12. Y. C. Hao *et al.*, *ACS Appl. Mater. Inter.* **9**, 40273 (2017).
13. N. N. Hu *et al.*, *ACS Appl. Mater. Inter.* **10**, 18665 (2018).
14. G. Y. Zheng *et al.*, *Nano Lett.* **13**, 1265 (2013).
15. Z. P. Zeng and X. B. Liu, *Adv. Mater. Inter.* **5**, 1701274 (2017).
16. A. Eftekhari and D. W. Kim, *J. Mater. Chem.* A **5**, 17734 (2017).
17. X. X. Gu and C. Lai, *J. Mater. Res.* **33**, 16 (2018).
18. S. H. Yu *et al.*, *Accounts Chem. Res.* **51**, 273 (2018).
19. D. H. Liu *et al.*, *Adv. Sci.* **5**, 1700270 (2018).
20. Z. Yuan *et al.*, *Nano Lett.* **16**, 519 (2015).
21. Q. Pang, D. Kundu and L. F. Nazar, *Mater. Horiz.* **3**, 130 (2016).
22. Z. Li *et al.*, *ACS Appl. Mater. Inter.* **9**, 38477 (2017).
23. G. M. Zhou *et al.*, *ACS Nano* **7**, 5367 (2013).
24. J. Pu *et al.*, *Nano Energy* **37**, 7 (2017).
25. J. Zhou *et al.*, *Electrochim. Acta* **218**, 243 (2016).
26. C. Ye *et al.*, *Adv. Funct. Mater.* **27**, 1702524 (2017).
27. G. M. Zhou *et al.*, *Proc. Natl. Acad. Sci. USA* **114**, 840 (2017).
28. M. Li *et al.*, Mater. *Res. Bull.* **96**, 509 (2017).
29. D. Q. He *et al.*, *J. Electrochem. Soc.* **164**, A1499 (2017).
30. Z. L. Ma *et al.*, *J. Power Sources* **325**, 71 (2016).
31. S. S. Zhang and D. T. Tran, *J. Mater. Chem.* A **4**, 4371 (2016).
32. Y. Lu *et al.*, *Nanoscale* **8**, 17616 (2016).
33. J. Park *et al.*, *Adv. Energy Mater.* **7**, 1602567 (2017).
34. X. L. Li *et al.*, *J. Alloys Compd.* **692**, 40 (2017).
35. Z. B. Cheng *et al.*, *Adv. Energy Mater.* **8**, 1702337 (2018).
36. Z. W. Seh *et al.*, *Nat. Commun.* **5**, 5017 (2014).
37. T. Chen *et al.*, *Nano Energy* **38**, 239 (2017).

38. W. Liu *et al.*, *Proc. Natl. Acad. Sci.* **114**, 3578 (2017).
39. T. Chen *et al.*, *Nano Energy* **38**, 239 (2017).
40. P. Q. Guo *et al.*, *Electrochim. Acta.* **256**, 28 (2017).
41. H. T. Wang *et al.*, *Nano Lett.* **14**, 7138 (2014).
42. K. Liang *et al.*, *Adv. Energy Mater. Lett.* 7, 1701309 (2017).
43. H. J. Peng and Q. Zhang, *Angew. Chem. Int. Ed.* **54**, 11018 (2015).
44. H. J. Peng *et al.*, *Angew. Chem. Int. Ed.* **55**, 12990 (2016).
45. C. J. Hart *et al.*, *Chem. Commun.* **51**, 2308 (2015).
46. K. Liang *et al.*, *Adv. Energy Mater.* 7, 1701309 (2017).
47. B. Jache *et al.*, *J. Power Sources* **247**, 703 (2014).
48. Q. Pang *et al.*, *Nat. Commun.* **5**, 4759 (2014).
49. Q. Zou and Y. C. Lu, *J. Phys. Chem. Lett.* 7, 1518 (2016).
50. C. J. Tang *et al.*, *Adv. Funct. Mater.* **28**, 1704561 (2018).
51. H. B. Lin *et al.*, *Energy Environ. Sci.* **10**, 1476 (2017).
52. H. L. Ye *et al.*, Proc. *Natl. Acad. Sci. USA* **114**, 13091 (2017).
53. T. S. Sonia *et al.*, *Ceram. Int.* **40**, 8351 (2014).
54. Y. S. Su and A. Manthiram, *J. Power Sources* **270**, 101 (2014).
55. J. B. Ye *et al.*, *Electrochim. Acta.* **190**, 538 (2016).
56. L. Ma *et al.*, *J. Mater. Chem.* A **3**, 19857 (2015).
57. D. Lin *et al.*, *Energy Environ. Sci.* **8**, 2371 (2015).
58. Z. H. Liu *et al.*, *Energy Storage Mater.* **13**, 112 (2018).
59. X. F. Yang *et al.*, *Electrochem. Energ. Rev.* **1**, 1 (2018).

Chapter 4

Graphene Oxide-Polypyrrole Composite as Sulfur Hosts for High-Performance Lithium–Sulfur Batteries

Qian Wang*, Chengkai Yang*, Hui Tang*, Kai Wu*, and Henghui Zhou*,†,‡

Lithium-sulfur batteries are considered as a promising candidate for the next-generation high energy density storage devices. However, they are still hindered by serious capacity decay on cycling caused by the dissolution of redox intermediates. Here, we designed a unique structure with polypyrrole (ppy) inserting into the graphene oxide (GO) sheet for accommodating sulfur. Such a sulfur host not only exhibits a good electronic and ionic conductivity, but also can suppress polysulfide dissolution effectively. With this advanced design, the composite cathode showed a high specific capacity of 548.4 mA·h·g^{-1} at 5.0 C. A stable Coulombic efficiency of ~99.5% and a capacity decay rate as low as 0.089% per cycle along with 300 cycles at 1.0 C were achieved for composite cathodes with 78 wt.% of S. Besides, the interaction mechanism between PPy and lithium polysulfides (LPS) was investigated by density-functional theory (DFT), suggesting that only the polymerization of N atoms can bind strongly to Li ions of LPS rather than single N atoms. The 3D structure GO-PPy host

*College of Chemistry and Molecular Engineering, Peking University, Beijing 100871, P. R. China.
†Beijing Engineering Research Center of Power Lithium-ion Battery, Beijing 102202, P. R. China.
‡hhzhou@pku.edu.cn

with high conductivity and excellent trapping ability to LPS offered a viable strategy to design high-performance cathodes for Li–S batteries.

Keywords: Li–S batteries; cathode; GO; polypyrrole; DFT calculation.

1. Introduction

Lithium-ion batteries (LIBs) have been widely used in electric vehicles, portable electronic products and grid-scale energy storage systems.[1-3] However, traditional LIBs-based graphite anode and transition metal oxide cathode can hardly meet people's demand for high energy density storage devices. It is imperative to develop next generation LIBs with higher energy density and longer cyclic stability.[4,5] Recently, rechargeable lithium–sulfur (Li–S) batteries have attracted great interest due to their high theoretical specific capacity (1675 mAh·g^{-1}) and high theoretical energy density (2600 Wh·kg^{-1}).[6-10] Beyond that, when compared with traditional cathode materials, sulfur is a low cost and environmental-friendly element. However, several major issues still limit the practical application of Li–S batteries, including: (1) the poor conductivity of S and its discharge products (Li$_2$S, Li$_2$S$_2$); (2) the low active material utilization caused by S and polysulfide dissolution; (3) dissolution of LPS into the electrolyte and shuttle effect; (4) the huge volume expansion (80%) during charging–discharging process.[11-14]

Over the last few decades, much effort has been devoted to solving these problems. The most common strategy is focused on confining LPS by using hosts to encapsulate sulfur, such as carbon nanotubes,[15-17] graphene,[18-20] hollow carbon sphere[21-25] and metal-organic frameworks[26,27] and so forth. Cui *et al.* demonstrated a yolk–shell TiO$_2$ host to accommodate sulfur, the obtained electrode put up an improved cycling stability at 0.5 C.[28] Nazar *et al.* proposed the carbon nanospheres host to encapsulate sulfur, and the sulfur cathode showed an enhanced cycle performance at 0.2 C and 0.5 C.[23] Landon *et al.* adopted the carbon nanotubes to store sulfur, and achieved an excellent electrochemical performance at 0.1 C with a sulfur mass loading of ~79 wt.%.[16] However, these hosts mentioned above still limit their effectiveness in high-rate and high-loading sulfur.

An ideal sulfur host should have the following properties: (1) high specific surface areas to accommodate high-loading S and improve the utilization of S; (2) good conductivity to ensure it can run at high rate with high specific capacity; (3) strong interaction with LPS to efficiently prevent polysulfide migration to the Li anode; (4) simple and low-cost synthetic route to provide a possibility for commercialization. Although a number of 3D host structures based on carbon spheres, carbon fibers, transition metal oxide, and polymers have been reported in the literature, they fall short in one or more of these aspects.

In this work, we proposed the 3D structure GO-PPy host by a simple self-assembly strategy with surfactant template. The GO-PPy composite host demonstrated an obvious 3D structure with PPy polymerizing in the GO sheet *in-situ.* Such a 3D structure with high conductivity and specific surface area not only improved the utilization of S, but also can effectively trap the intermediate polysulfides. With that, the GO-PPy-S electrode exhibited improved cycling stability and excellent rate capability. With a sulfur content of 78 wt.% in the cathode material, GO-PPy electrode demonstrated a high initial capacity of 848.3 $mA·h·g^{-1}$ and a capacity decay rate as low as 0.089% per cycle along with 300 cycles at 1.0 C. Besides, the composite cathode displayed a high specific capacity of 548.4 $mA·h·g^{-1}$ at 5.0 C. The interaction mechanism between lithium polysulfides and PPy was investigated by DFT, and calculation results indicated that only the PPy molecule with polymerization of N atoms can bind strongly to LPS rather than pyrrole with single N atom. Thus, we anticipate that the GO-PPy host with high conductivity and strong binding ability to LPS provides an effective strategy for the development of high-performance Li–S batteries.

2. Materials and Methods

The approach is based on the electrostatic interaction theory. It is well known that the GO sheets are negatively charged. In an aqueous solution, cationic surfactant of CTAB can be self-assembled into a rod micelle and combined with graphene layers by the electrostatic interaction when the concentration is far above its second critical micellization concentration (CMC). After introducing pyrrole monomer into the mixed solution, the pyrrole molecule would tend to solubilize in the hydrophobic cavity of the CTAB micelles due to the similarity compatibility principle.[29] Then, adding the initiator of ammonium persulfate (APS), polymerization would also take place in the hydrophobic micelle cavity predominately. At last, a 3D structure GO-PPy composite was achieved after removing the surfactant template by washing.

3. Results and Discussion

As shown in Fig. 1(a), obvious layer structure was observed on the surface of GO sample. After polymerization, the PPy fibers either sandwiched between GO sheets or on the GO surface (Fig. 1(b)), and the PPy fibers were interlaced between GO sheets. Such a GO-PPy composite structure combined the unique properties of GO and PPy, such as 3D carbon matrix, high specific surface areas, good conductivity etc. It could be an ideal host for S element. After sulfur infiltration, the sulfur was well encapsulated inside the GO sheets and coated on the surface of PPy fibers with nodulelike particles ~2 μm (Fig. 1(c)).

In order to further understand the GO-PPy composite, FTIR characterizations were conducted (see supplementary material). For pure GO, a wider and stronger absorption peak was detected at around 3400 cm^{-1}, which can be assigned to the vibration of OH. The peaks at 1726 cm^{-1}, 1390 cm^{-1} were due to vibrations of C = O

Fig. 1.   SEM images of (a) GO, (b) GO-PPy, (c) GO-PPy-S.

and C–OH. Meanwhile, vibrations at 1624 cm^{-1}, 1227 cm^{-1}, and 1057 cm^{-1} were attributed to the bands of C = C, C–O and C–O, respectively. The results indicated that there were a lot of functional groups in the surface of GO, which could trap LPS to some extent. And for the spectra of PPy, the peaks at 1553 cm^{-1} and 1475 cm^{-1} were confirmed in the presence of pyrrole rings. The absorption peaks at 1383 cm^{-1}, 1192 cm^{-1} were caused by vibrations of C–N and C = N, respectively. The peaks in the range of 500–750 cm^{-1} and at 1045 cm^{-1} were originated from C–H vibration-absorption. The absorption at 793 cm^{-1} could be attributed to the out-of-plane bending vibration of N–H. Notably, all the peaks of GO and PPy could be seen from GO-PPy composites. Furthermore, no excess peaks were detected, which suggested that the CTAB had been removed completely.

Figure S2 (see supplementary material) exhibited the XPS analysis of the GO, GO-PPy and GO-PPy-S composite. Peak positions of O 1s was located at 532 eV, which were detected in GO sample. This suggested the presence of oxygen functional groups on the surface of GO, which was consistent with the spectra of FTIR. From the C 1s XPS, the peak at 283.8 eV was observed in GO-PPy, which can be assigned as the C = C band in PPy molecule. Combining the N 1s spectra of GO-PPy, the peak at ~400 eV was related to the pyrrole-like nitrogen (pyrrolic-N), indicating that the PPy was successfully polymerized in GO sheets. Comparing the peak position of N 1s and O 1s in three samples, it could be found that all the N 1s, O 1s peaks were located at 400 eV and 532 eV, respectively, which indicated that there has been no interaction between S and GO-PPy composite and the chemical form of S is elemental sulfur. For providing further evidence, XRD data was shown in Fig. S3 (supplementary material). The patterns appeared similar in the three samples, which suggested that sulfur was immobilized within the GO-S, GO-PPy-S composites and there was no chemical reaction between sulfur and GO-PPy.

After loaded with sulfur, the mass loading of sulfur in the GO-S and GO-PPy-S composites was measured by TG, showing a high loading content of 78 wt.%, which could promise a sulfur cathode material with high energy density for Li–S batteries (Fig. S4(a), supplementary material). Adsorption ability was compared in between by adding 10 mg GO and GO-PPy in 2 mL of 10 mM Li_2S_6 in DOL/ DME, which was shown in the inset of Fig. S4(b). The GO-PPy composite displayed a strong trapping ability for LPS, and the solution became clear and colorless after 5 h. However, the solution with adding pure GO only performed a little faded after 5 h, indicating extremely weak interaction between PPy and LPS. Besides, electrochemical impedance spectrum (EIS) measurements were carried out to compare the GO-S and GO-PPy-S electrodes. As shown in Fig. S4(b), a high interfacial resistance (~160 Ω) was detected with GO-S electrode. However, the interfacial resistance for GO-PPy-S electrode exhibited a lower value of ~70 Ω, indicating the conductivity of the GO-PPy-S cathode was improved over the pure GO-S cathode. Thus, the GO-PPy host with strong binding ability to LPS and improved conductivity could ensure a highperformance cathode material for Li–S batteries compared to bare GO.

The electrochemical properties of GO-S and GO-PPy-S electrodes was evaluated in a coin cell with a Li metal anode. The cell was first discharged and charged at C-rates of 0.1 to 5.0 C, and then tested for cycling stability at 1.0 C. As shown in Fig. 2(a), the GO-PPy-S electrode delivered a high specific discharge capacity of 1153 mA·h·g^{-1} at 0.2 C. Further cycling at 0.5 C, 1.0 C and 2.0 C, the electrode showed high reversible capacities of 865.1mA·h·g^{-1}, 758.6 mA·h·g^{-1} and 676.5 mA·h·g^{-1}, respectively. Notably, when the C-rate was increased to 5.0 C, a high discharge capacity of 548.4 mA·h·g^{-1} was observed, indicating the excellent electronic and ionic transport properties of the GO-PPy-S electrode. When the current density was reduced back to 1.0 C again, the discharge capacity of the GO-PPy-S cell was recovered to 748.6 mA·h·g^{-1}, manifesting the excellent stability of the cathode structure. In contrast, the GO-S electrode demonstrated an inferior capacity to that of GO-PPy-S cell, especially at high-rate. Besides, from the charging/discharging profiles at various rates (Fig. 2(b)), the two-plateau characteristic can be clearly identified from 0.2 C to 5.0 C, suggesting a low polarization and high electrical conductivity in GO-PPy-S electrode. Such an improved rate of performance could be attributed to the good electronic, ionic conductivity and excellent binding ability with LPS of GO-PPy composite material, which was

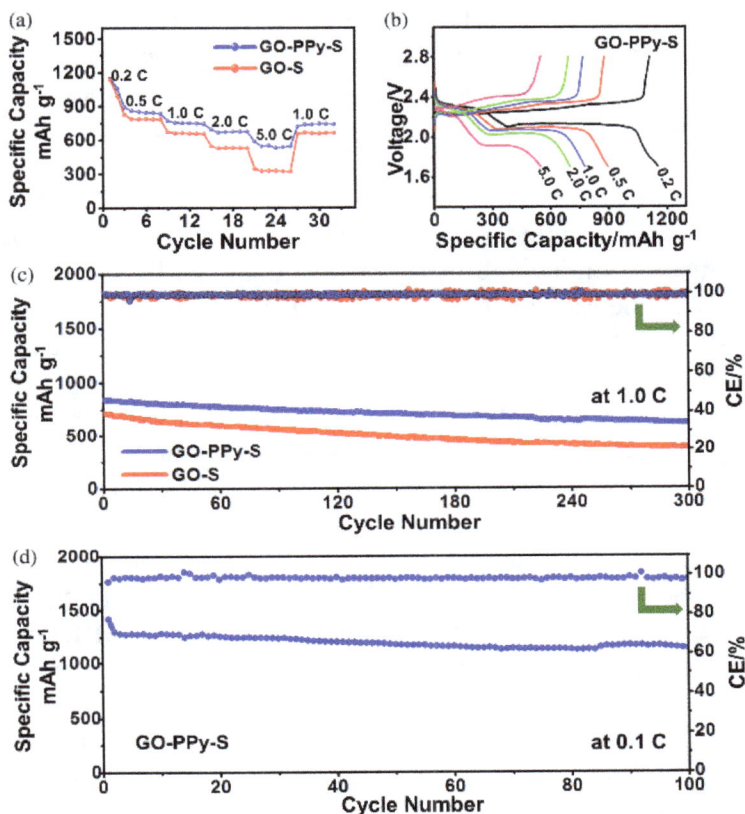

Fig. 2. (a) The discharging capacities of GO-PPy-S and GO-S electrodes cycled at various C-rates from 0.2 C to 5.0 C. (b) Representative charging–discharging voltage profiles of GO-PPy-S electrode at different C rates. (c) Cycle performance of GO-Ppy-S and GO-S cathodes with 1.2 mg·cm^{-2} sulfur mass loading at 1.0 C. (d) Cycling stability of GO-Ppy-S cathode with high loading (~3:0 mg·cm^{-2}) at 0.1 C.

consistent with the results of EIS. Figure 2(c) showed the cycling performance of the cells at 1.0 C in a potential range from 1.7 V to 2.8 V. The GO-PPy-S electrode displayed a high initial capacity of 848.3 mA·h·g^{-1}, and it retained a capacity of 621.9 mA·h·g^{-1} after 300 cycles, corresponding to an average capacity loss of 0.089% per cycle. In comparison, the pure GO-S electrode exhibited a drastic capacity decay and a low capacity of 385.7 mA·h·g^{-1} after 300 cycles.

We further increased the sulfur mass on the GO-PPy-S electrode to ~3.0 mg·cm⁻². As shown in Fig. 2(d), a high specific capacity of about 1400 mA·h·g⁻¹ was achieved at 0.1 C for higher mass loading. Furthermore, the electrodes also showed a very high Coulombic efficiency (>98%) and good cycling stability within 100 cycles. The excellent cycling performance of the GO-PPy-S electrode with high loading S was attributed to the unique structure of GO-PPy host.

To further confirm the strong trapping ability of GO-PPy-S cathode to LPS, the cells with GO-S and GO-PPy-S cathodes were disassembled after 300 cycles, which was shown in Fig. S5 (supplementary material). Compared to the photograph of the disassembled cells, the separator of cell with GO-S cathode exhibited an obvious yellow color (Fig. S5(a)), because it contained a significant amount of LPS, whereas, the separator with GO-PPy-S cathode was clear and almost free of LPS (Fig. S5(b)). Besides, comparing the SEM image of used Li and EDX mapping, the cycled Li anode paired with the GO-S cathode exhibited a lot of cracks and breaks, while that paired with GO-PPy-S cathode showed as complete and relatively smooth. Combining the EDX mapping, a lower content of sulfur and carbon species was detected on the surface of Li metal paired with GO-PPy-S cathode (Fig. S5(c) and S5(d)). Thus, the GO-PPy host had a strong trapping ability to LPS and shuttle effect can be substantially suppressed in the long cycle.

The LPS binding mechanism was also explored by DFT calculation (Fig. 3). The binding free energy (ΔG_B) was employed to evaluate

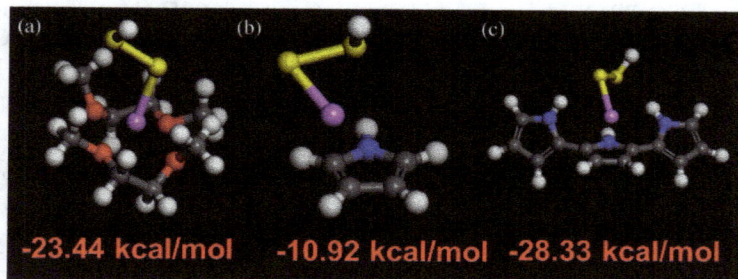

Fig. 3.   DFT calculations of binding free energies (ΔG_B). (a) Optimized geometries for the binding of Li-SS-H molecule to two DME molecules; (b) a pyrrole molecule; (c) three pyrrole molecule polymer.

the interaction strength between LiSSH and solvent molecule or binding molecule. From the previous report, LiSSH molecule prefers to be solvated by the chelating ligand DME rather than DOL.[30] Here, we only focused on the ΔG_B between DME and LiSSH, and found that the ΔG_B for binding of LiSSH to a solvent molecule DME was −19.89 kcal/mol (Fig. S6, supplementary material). However, for binding of LiSSH to two DME molecules, the ΔG_B was −23.44 kcal/mol. Therefore, LiSSH prefers to be solvated by two DME molecules in electrolyte. Then, we turn to calculate the ΔG_B for binding of LiSSH to pyrrole (Py) monomer and PPy. As shown in Fig. 3(b), the ΔG_B for binding of LiSSH to a Py molecule was −10.92 kcal/mol. However, for polymer of three Py molecules (Py-3), the ΔG_B was −28.33 kcal/mol (Fig. 3(c)). Hence, it is thermodynamically favorable for the Py-3 to replace DME molecule and bind LPS. Put this and that together, the binding ability of LiSSH with PPy would be very strong, which is beneficial for suppressing dissolution of LPS and shuttle effect.

Our calculation results indicated that single N atom cannot suppress shuttle effect and dissolution of LPS because of its weak interaction with LPS. However, PPy displayed a strong interaction with LPS, which was consistent with the adsorption tests. The calculations offered a theoretical basis for designing effective hosts for high-performance Li–S batteries.

4. Conclusion

In summary, we have designed a unique GO-PPy composite structure as sulfur host for high-performance Li–S batteries. Such a unique structure with PPy inserting into the GO sheet exhibits good electronic and ionic conductivity. Besides, the PPy with multiple N atoms provides strong affinity to polysulfides, which suppresses the dissolution of polysulfides and shuttle effect effectively. With this advanced design, the GO-PPy-S electrode with high sulfur loading content of 78 wt.% displays much enhanced electrochemical performance, including CE, rate capability and cycling stability. Thus, we believe that this design strategy and DFT calculations may guide us in developing high-performance sulfur host for Li–S batteries.

Acknowledgments

PULEAD Technology Industry Co. Ltd and the National Natural Science Foundation of China (21875004) are gratefully acknowledged for their financial support for this work. Qian Wang, Chengkai Yang and Hui Tang contributed equally to this work.

References

1. J. W. Choi and D. Aurbach, *Nat. Rev. Mater.* **1**, 16013 (2016).
2. A. Fotouhi, D. J. Auger, K. Propp, S. Longo and M. Wild, Renew. *Sust. Energ. Rev.* **56**, 1008 (2016).
3. J. Yang, Z. Li, J. Wang, Q. Xiao and G. Lei, *Funct. Mater. Lett.* **5**, 1250019 (2012).
4. P. G. Bruce, S. A. Freunberger, L. J. Hardwick and J. M. Tarascon, *Nat. Mater.* **11**, 19 (2012).
5. A. Manthiram, Y. Z. Fu, S. H. Chung, C. X. Zu and Y. S. Su, *Chem. Rev.* **114**, 11751 (2014).
6. A. Manthiram, Y. Z. Fu and Y. S. Su, *Accounts Chem. Res.* **46**, 1125 (2013).
7. Y. Hao, D. Xiong, W. Liu, L. Fan, D. Li and X. Li, ACS *Appl. Mater. Interfaces* **9**, 40273 (2017).
8. N. Hu, X. Lv, Y. Dai, L. Fan, D. Xiong and X. Li, ACS *Appl. Mater Interfaces* **10**, 18665 (2018).
9. S. Y. Bai, X. Z. Liu, K. Zhu, S. C. Wu and H. S. Zhou, *Nat. Energy* **1**, 16094 (2016).
10. H. Cheng, S. Wang, D. Tao and M. Wang, *Funct. Mater. Lett.* **7**, 1450020 (2014).
11. Y. Zhao, M. Liu, W. Lv, Y. B. He, C. Wang, Q. B. Yun, B. H. Li, F. Y. Kang and Q. H. Yang, *Nano Energy* **30**, 1 (2016).
12. Z. H. Sun, J. Q. Zhang, L. C. Yin, G. J. Hu, R. P. Fang, H. M. Cheng and F. Li, *Nat. Commun.* **8**, 14627 (2017).
13. B. Yan, X. Li, Z. Bai, X. Song, D. Xiong, M. Zhao, D. Li and S. Lu, J. *Power Sources* **338**, 34 (2017).
14. G. M. Zhou, H. Z. Tian, Y. Jin, X. Y. Tao, B. F. Liu, R. F. Zhang, Z. W. Seh, D. Zhuo, Y. Y. Liu, J. Sun, J. Zhao, C. X. Zu, D. S. Wu, Q. F. Zhang and Y. Cui, P Natl. *Acad. Sci.* **114**, 840 (2017).
15. S. H. Chung, R. Singhal, V. Kalra and A. Manthiram, *J. Phy. Chem. Lett.* **6**, 2163 (2015).

16. M. Y. Li, R. Carter, A. Douglas, L. Oakes and C. L. Pint, *ACS Nano* **11**, 4877 (2017).
17. D. J. Xiao, H. F. Zhang, C. M. Chen, Y. D. Liu, S. X. Yuan and C. X. Lu, *Chemelectrochem.* **4**, 2959 (2017).
18. M. P. Yu, R. Li, M. M. Wu and G. Q. Shi, *Energy Storage Mater.* **1**, 51 (2015).
19. G. M. Zhou, L. Li, D. W. Wang, X. Y. Shan, S. F. Pei, F. Li and H. M. Cheng, *Adv. Mater.* **27**, 641 (2015).
20. D. W. Su, M. Cortie and G. X. Wang, Adv. *Energy Mater.* **7**, 1602014 (2017).
21. N. Jayaprakash, J. Shen, S. S. Moganty, A. Corona and L. A. Archer, Angew. *Chem. Int. Ed.* **50**, 5904 (2011).
22. C. F. Zhang, H. B. Wu, C. Z. Yuan, Z. P. Guo and X. W. Lou, *Angew. Chem. Int. Ed.* **51**, 9592 (2012).
23. G. He, S. Evers, X. Liang, M. Cuisinier, A. Garsuch and L. F. Nazar, *ACS Nano* **7**, 10920 (2013).
24. W. D. Zhou, Y. C. Yu, H. Chen, F. J. DiSalvo and H. D. Abruna, *J. Am Chem Soc.* **135**, 16736 (2013).
25. G. M. Zhou, Y. B. Zhao and A. Manthiram, *Adv. Energy Mater.* **5** (2015).
26. Z. Liang, C. Qu, W. Guo, R. Zou and Q. Xu, *Adv Mater.* (2017).
27. D. J. Xiao, Q. Li, H. F. Zhang, Y. Y. Ma, C. X. Lu, C. M. Chen, Y. D. Liu and S. X. Yuan, *J. Mater. Chem.* A **5**, 24901 (2017).
28. Z. Wei Seh, W. Li, J. J. Cha, G. Zheng, Y. Yang, M. T. McDowell, P. C. Hsu and Y. Cui, *Nat. Commun.* **4**, 1331 (2013).
29. L. L. Zhang, S. Zhao, X. N. Tian and X. S. Zhao, *Langmuir.* **26**, 17624 (2010).
30. W. Liu, J. Jiang, K. R. Yang, Y. Mi, P. Kumaravadivel, Y. Zhong, Q. Fan, Z. Weng, Z. Wu, J. J. Cha, H. Zhou, V. S. Batista, G. W. Brudvig and H. Wang, *Proc. Natl. Acad. Sci.* **114**, 3578 (2017).

Chapter 5

Synthesis of Carbon Nanoflake/Sulfur Arrays as Cathode Materials of Lithium–Sulfur Batteries

Fan Wang*, Xinqi Liang*, Minghua Chen*,§, and Xinhui Xia[†,‡]

It is of great importance to develop high-quality carbon/sulfur cathode for lithium-sulfur batteries (LSBs). Herein, we report a facile strategy to embed sulfur into interconnected carbon nanoflake matrix forming integrated electrode. Interlinked carbon nanoflakes have dual roles not only as a highly conductive matrix to host sulfur, but also act as blocking barriers to suppress the shuttle effect of intermediate polysulfides. In the light of these positive characteristics, the obtained carbon nanoflake/S cathode exhibits good LSBs performances with high capacities (1117 mAh g^{-1} at 0.2°C, and 741 mAh g^{-1} at 0.6°C) and good high-rate cycling performance. Our synthetic method provides a novel way to construct enhanced carbon/sulfur cathode for LSBs.

*Key Laboratory of Engineering Dielectric and Applications (Ministry of Education), Harbin University of Science and Technology, Harbin 150080, P. R. China.
†State Key Laboratory of Silicon Materials, Key Laboratory of Advanced Materials and Applications for Batteries of Zhejiang Province, and Department of Materials Science and Engineering, Zhejiang University, Hangzhou 310027, P. R. China.
‡Institute of Advanced Electrochemical Energy, Xi'an University of Technology, Xi'an 710048, P. R. China.
§chenminghuahrb@126.com

Keywords: Lithium sulfur batteries; carbon nanoplate; hydrothermal method; electrochemical energy storage; arrays.

1. Introduction

The growing crisis of fossil fuel and environment issues make it urgent to develop advanced batteries with high energy and power density, and stable cycling life.[1-4] Nowadays, lithium ion batteries (LIBs) have been widely used in portable electronics, however, their energy density is too low to meet the demands of electric transportation.[5-8] Hence, lithium-sulfur batteries (LSBs) appear to be one of the most attractive candidates for the high-energy lithium batteries due to their high theoretical specific capacity of about 1675 mAh g^{-1}, much larger than the 175–250 mAh g^{-1} of conventional positive electrode materials.[9-12] Furthermore, the abundance and non-toxicity advantages of sulfur make LSBs more economy and environmental friendly.

Unfortunately, several major drawbacks of sulfur cathode have hindered its large-scale applications. First, sulfur is both ionically and electrically insulating, which leads to a poor utilization of active material. Second, shuttle effect of intermediate polysulfides is another serious problem and causes loss of active materials.[13] The third problem arises from the volume variation of the sulfur cathode upon cycling. Due to different densities of S_8 (2.07 g cm^{-3}) and Li_2S (1.66 g cm^{-3}),[14] sulfur exhibits a volume expansion of about 79% upon lithiation.[15-18] Consequently, considerable efforts are dedicated to solving these problems. One of the most efficient methods is to pump sulfur into carbon nanostructure matrix to reinforce the electrical conductivity of the whole electrode, release volume expansion and slow down the shuttle effect. Until now, numerous nanostructure carbon matrix materials, such as porous carbon,[1,2] graphene or grapheme oxide,[19,20] and carbon nanotube/nanofiber,[21,22] have been synthesized and utilized to improve the electrochemical performance of sulfur cathode. For example, Zhong *et al.*[23] reported porous rice carbons and porous carbon fibers as the conductive matrix for sulfur and showed excellent capacity cyclability at a high discharge current rate. However, the above system is powder-form material and should be mixed with insulating binders, resulting in undermined performance. In such a context, integrated carbon/S cathode emerges due to its binder-free characteristics and good highrate properties. Typically, free-standing carbon nanoflake arrays are good matrixes for sulfur, but there is no work on their rational combination.

In this work, we report a self-supported carbon nanoflake/S arrays by a facile hydrothermal method. S is uniformly embedded into the carbon nanoflake matrix forming binderfree electrode. Arising from good conductive network, effective blocking effect for poly-sulfides, the designed carbon nanoflake/sulfur arrays show a high capacity of 1117 mAh g^{-1} at 0.2°C and good cycles. Our electrode protocol can provide reference for construction of other advanced sulfur cathode for LSBs.

2. Materials and Methods

The carbon nanoflakes matrix was fabricated by a hydrothermal-carbonization method as follows: Typically, glucose (0.6 g), urea (0.6 g) and $Zn(NO_3)_2$ (0.6 g) were mixed in 100 mL H_2O to form the hydrothermal solution. Then, this hydrothermal solution was put into Teflon liner and maintained at 200°C for 6 h. Nickel foam was put into the Teflon liner to serve as the substrate. After hydrothermal reaction, the samples were taken out and washed. Then, the samples were annealed at 650°C for 2 h in Argon to obtain carbon nanoflake arrays after immersing into 1M KOH for 2 h. After that, CS_2 solution with sulfur was dipped onto the above carbon nanoflake arrays. Then, the samples were dried at 50°C to remove CS_2. Finally, the sample was kept at 155°C for 12 h to form the final carbon nanoflakes/sulfur arrays cathode. The mass loading of sulfur was about 2.5 mg cm^{-2}.

Morphologies and structures of samples were characterized by field emission scanning electron microscopy (FESEM, SU-70), Renishaw Raman microscope under 532 nm laser excitation, transmission electron microscopy (TEM, JEOLJEM200CX), and X-ray diffraction (XRD, RIGAKUD/Max-2550 with Cu Kα radiation). The mass of sulfur was measured by Thermogravimetric Analysis (TGA, Mettler ToledoSDTQ600).

The carbon nanoflakes/sulfur arrays were directly used as the working electrode. Lithium metal was used as the counter electrode and reference electrode. A polypropylene microporous film (Cellgard 2300) was used as the separator. The electrolyte consisted of 1 M bis (trifluoromethane) sulphonamide lithium salt (LiTFSI) in a mixed solvent of 1,3-dioxolane (DOL) and 1,2-dimethoxyethane (DME) with a volume ratio of 1:1, including 1 wt.% $LiNO_3$ as an electrolyte additive. Cyclic voltammograms (CVs) were conducted at electro-chemical workstation (CHI 660D) in the potential range of 1.5–3.0 V (Li/Li$^+$) at a scan rate of 0.1 mVs^{-1}. The discharge/charge test was conducted on a NewWei battery system between 1.6 and 2.7 V (Li/Li$^+$) at room temperature.

3. Results and Discussion

SEM images of carbon nanoflake and carbon nanoflake/S arrays are shown in Fig. 1. After hydrothermal growth of carbon nanoflakes,

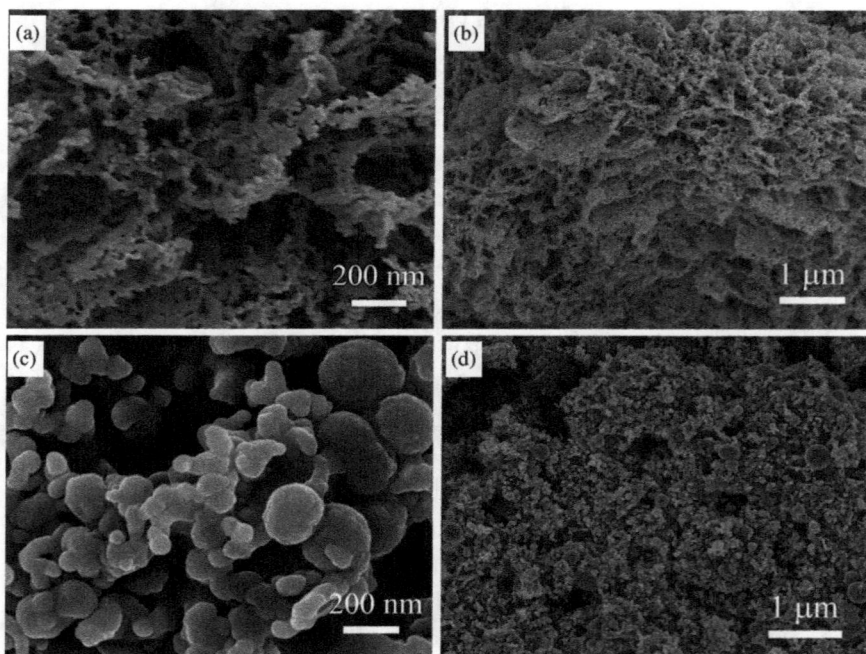

Fig. 1. SEM images of (a, b) carbon nanoflakes arrays and (c, d) carbon nanoflake/S arrays.

note that self-supported carbon nanoflakes show highly porous structure and are composed of interlinked nanoparticles of 20–100 nm (Figures 1(a) and 1(b)). The surface of carbon nanoflakes exhibit high roughness. It should be mentioned that the highly porous structure of carbon nanoflake matrix is favorable for sufficient soaking of electrolyte and provides short diffusion paths of ions/electrons. In addition, its interlinked network can form labyrinth to stop the shuttle of polysulfides. To utilize these positive advantages, we embed sulfur into the above carbon nanoflake matrix forming integrated arrays electrode. After embedding sulfur, the hierarchical porous architecture of the carbon nanoflake/S arrays is well kept, but the surface morphology of the carbon nanoflake/S arrays becomes smooth (Figs. 1(c) and 1(d)). The composite particle size becomes a bit larger up to 30–200 nm. Obviously, the sulfur is homogeneously embedded into the carbon nanoflake arrays. Moreover, the integrated porous architecture will not block the transfer of electrolyte.

As presented in TEM images (Fig. 2(a)), the flake texture of carbon matrix is well noticed and it consists of interconnected nanoparticles. Selected area electron diffraction (SAED) proves that the obtained carbon nanoflake shows amorphous nature. After compositing with sulfur, the smooth surface of carbon nanoflake/S is detected (Fig. 2(b)), indicating that the sulfur is completely embedded into the carbon nanoflakes, verified by the EDS elemental mapping images of C and S (Fig. 2(c)). Comparing the XRD patterns of carbon nanoflakes and carbon nanoflake/S arrays in Figs. 3(a) and 3(b), it is seen that except the substrate peaks of nickel foam, the carbon nanoflakes/S composite electrode contains characteristic peaks of C and S (Fig. 3(b)).[24] The sulfur shows orthorhombic structure (JCPDS 08-0247), suggesting that the anchor of sulfur in carbon nanoflakes is a physical effect and keeps its original state. The sulfur in the carbon nanoflake is confirmed by Raman measurement (Fig. 3(c)). Three peaks of sulfur (153 cm^{-1}, 220 cm^{-1} and 472 cm^{-1}) (Fig. 3(c)) are noticed except for the carbon peaks from carbon nanoflakes matrix (1350 cm^{-1} and 1585 cm^{-1}).[25] It is indicated that the designed carbon nanoflakes are good matrixes for accommodation of sulfur to form

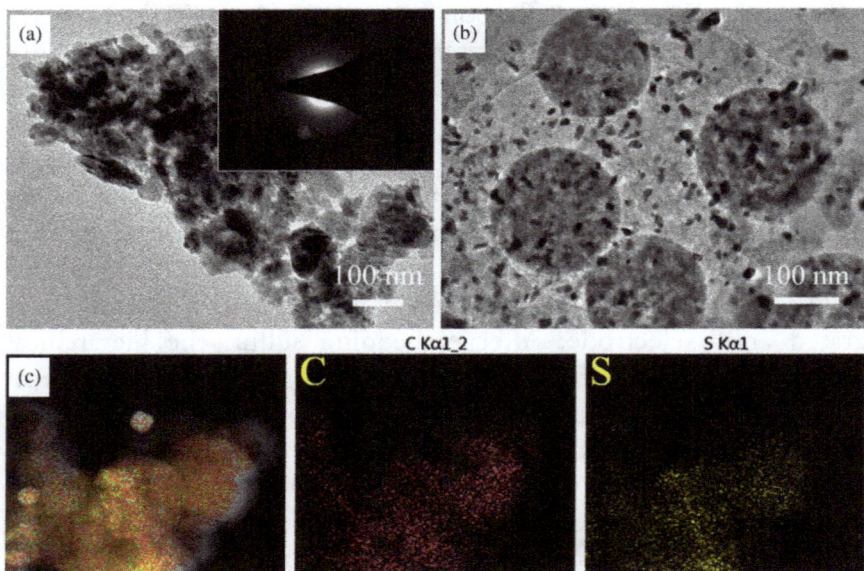

Fig. 2. TEM images of (a) carbon nanoflake (SAED pattern in inset) and (b) carbon nanoflake/S nanoflakes and (c) EDS elemental mapping images of C and S.

integrated carbon nanoflakes/S cathode. According to the TG result (Fig. 3(d)), the load mass of sulfur is about 67.7%.

The carbon nanoflakes/S arrays electrodes were characterized as cathodes of LSBs. Figure 4(a) presents CV curves at the second cycle at a scan rate of 0.1 mVs^{-1}. Characteristic CV behavior of sulfur cathode is noticed. During the cathodic process, the peak at 2.32 V is due to the conversion from S to long-chain polysulfides Li_2Sn ($4 \leq n \leq 8$). The peak at 2.02 V is owing to change from long-chain polysulfides to short-chain polysulfides Li_2Sn ($n < 4$) and finally to Li_2S.[1,2] During the anodic process, the peak at 2.33 V is owing to the conversion from Li_2S to polysulfides. The peak at 2.41 V is due to the reformation of S.

The discharge/charge curves at different cycles and rate performance are presented in Figs. 3(b) and 3(c). It is seen that typical charge/discharge plateaus owing to S cathode are detected. The carbon nanoflakes/S arrays electrode exhibits a high capacity of 1117 mAh g^{-1} at 0.2°C, 854 mAh g^{-1} at 0.4°C, 741 mAh g^{-1} at 0.6°C,

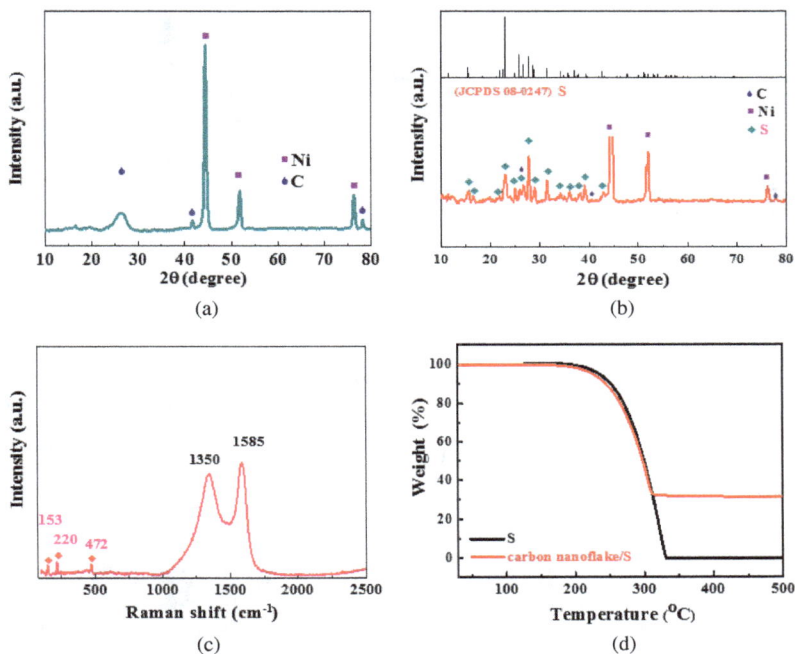

Fig. 3. XRD patterns of (a) carbon nanoflakes arrays and (b) carbon nanoflake/S arrays grown on nickel foam. (c) Raman spectrum of carbon nanoflake/S arrays. (d) TG curve of carbon nanoflake/S arrays.

632 mAh g^{-1} at 1.0°C, 533 mAh g^{-1} at 2.0°C, and 333 mAh g^{-1} at 5.0°C, respectively. Furthermore, the specific capacity is recovered to 917 mAh g^{-1} at 0.2°C. All these results demonstrate the carbon nanoflakes/S electrodes are good cathodes for LSBs. The cycling life of carbon nanoflakes/S composite electrodes at 0.4°C is shown in Fig. 4(d). Good stability is noticed for the carbon nanoflakes/S arrays electrode. After 200 cycles, the carbon nanoflakes/S arrays electrodes can deliver a specific capacity of 637 mAh g^{-1} at 0.4°C, with a low decay rate with only 0.125% per cycle at 0.4°C, indicating that the carbon nanoflakes/S arrays show effective blocking function to suppress the "shuttle effect" of polysulfides to hold a stable life. The following positive factors are responsible for the good performance. The carbon nanoflakes can provide good conductive hosts to

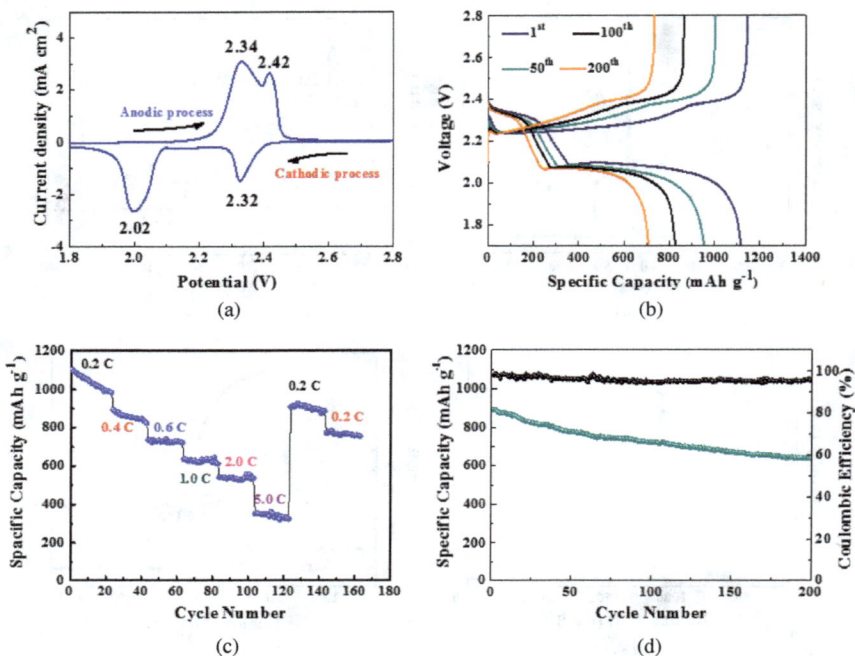

Fig. 4. (a) CV curve of carbon nanoflake/S arrays at a scan rate of 0.1 mVs⁻¹ at the second cycle. (b) Charge/discharge curves at different cycles at 0.2°C. (c) Rate performance of carbon nanoflake/S arrays. (d) Cycling life of carbon nanoflake/S arrays at 0.4°C.

accommodate sulfur. In addition, the carbon nanoflakes act as carbon barriers to suppress the shuttle of polysulfides to obtain good cycling performance.[26–29] Meanwhile, the porous architecture and effective combination between the sulfur and carbon nanoflakes can accelerate the transfer of electrons/ions.[30–32]

4. Conclusion

In summary, we have fabricated carbon nanoflakes/S arrays by a facile hydrothermal carbonization method. Carbon nanoflakes serve as hosts and backbones for sulfur. The carbon nanoflakes/S arrays cathode exhibits good electrochemical LSBs performances with a high capacity of 1117 mAh g⁻¹ and good cycles due to carbon nanoflakes, which possess high electrical conductivity and good hindering effect for polysulfides. Our works show a new way for fabrication of integrated binder-free sulfur cathode for LSBs.

Acknowledgments

The authors acknowledge financial support from Natural Science Foundation of China (51502063 and 51502263), Key Laboratory of Engineering Dielectrics and Its Application (Harbin University of Science and Technology), Ministry of Education (KF20171101), University Nursing Program for Young Scholars with Creative Talents in Heilongjiang Province (UNPYSCT-2015038), China Postdoctoral Science Foundation (2016T90306 and 2015M570301), Natural Science Foundation (E2015064) and Postdoctoral Science Foundation (LBH-TZ0615 and LBH-Z14120) of Heilongjiang Province of China, and Science Funds for Young Innovative Talents of HUST (201505).

References

1. Y. Zhong, X. Xia, S. Deng, J. Zhan, R. Fang, Y. Xia, X. Wang, Q. Zhang and J. Tu, *Adv. Energy Mater.* **8**, 1701110 (2018).
2. Y. Zhong, D. Chao, S. Deng, J. Zhan, R. Y. Fang, Y. Xia, Y. Wang, X. Wang, X. Xia and J. Tu, *Adv. Funct. Mater.* 1706391 (2018), DOI: 10.1002/adfm.201706391.
3. X. Xia, J. Zhan, Y. Zhong, X. Wang, J. Tu and H. J. Fan, *Small* **13**, 1602742 (2017).
4. S. Liu, X. Xia, Y. Zhong, S. Deng, Z. Yao, L. Zhang, X. B. Cheng, X. Wang, Q. Zhang and *J.* Tu, *Adv. Energy Mater.* **8**, 1702322 (2018).
5. Y. Pan, *Funct. Mater. Lett.* **10**, 1750067 (2017).
6. C. Chen, C. H. Zhao, Z. B. Hu and K. Y. Liu, *Funct. Mater. Lett.* **10**, 1650074 (2017).
7. W. Hu, T. Shen, H. Y. Hou, G. Y. Gan, B. J. Zheng, F. X. Li and J. H. Yi, *Funct. Mater. Lett.* **9**, 1650015 (2016).
8. H. Cheng, S. P. Wang, D. Tao and M. Wang, *Funct. Mater. Lett.* **7**, 1450020 (2014).
9. Q. Q. Xiong, J. J. Lou, X. J. Teng, X. X. Lu, S. Y. Liu, H. Z. Chi and Z. G. Ji, *J. Alloys Compd.* **743**, 377 (2018).

10. Q. Q. Xiong, H. Y. Qin, H. Z. Chi and Z. G. Ji, *J. Alloys Compd.* **685**, 15 (2016).
11. Q. Q. Xiong and Z. G. Ji, *J. Alloys Compd.* **673**, 215 (2016).
12. Q. Xiong, H. Chi, J. Zhang and J. Tu, *J. Alloys Compd.* **688** (Part B), 729 (2016).
13. Z. Sun, J. Zhang, L. Yin, G. Hu, R. Fang, H. M. Cheng and F. Li, *Nat. Commun.* **8**, 14627 (2017).
14. Y. X. Yin, S. Xin, Y. G. Guo and L. J. Wan, Angew. *Chem. Int. Ed.* **52**, 13186 (2013).
15. D. Bresser, S. Passerini and B. Scrosati, *Chem. Commun.* **49**, 10545 (2013).
16. A. Manthiram, *J. Phys. Chem. Lett.* **2**, 176 (2011).
17. D.-W. Wang, Q. Zeng, G. Zhou, L. Yin, F. Li, H.-M. Cheng, I. R. Gentle and G. Q. M. Lu, *J. Mater. Chem. A* **1**, 9382 (2013).
18. Y. Yang, G. Zheng and Y. Cui, *Chem. Soc. Rev.* **42**, 3018 (2013).
19. M.-K. Song, Y. Zhang and E. J. Cairns, *Nano Lett.* **13**, 5891 (2013).
20. C. Zu and A. Manthiram, *Adv. Energy Mater.* **3**, 1008 (2013).
21. W. Ahn, K.-B. Kim, K.-N. Jung, K.-H. Shin and C.-S. Jin, *J. Power Sources* **202**, 394 (2012).
22. G. Zheng, Y. Yang, J. J. Cha, S. S. Hong and Y. Cui, *Nano Lett.* **11**, 4462 (2011).
23. C. Liang, N. J. Dudney and J. Y. Howe, *Chem. Mater.* **21**, 4724 (2009).
24. X. Q. Niu, X. L. Wang, D. Xie, D. H. Wang, Y. D. Zhang, Y. Li, T. Yu and J. P. Tu, *ACS Appl. Mater. Interfaces* **7**, 16715 (2015).
25. C. X. Zu, N. Azimi, Z. C. Zhang and A. Manthiram, *J. Mater. Chem. A* **3**, 14864 (2015).
26. X. Zhao, Z. Zhao, M. Yang, H. Xia, T. Yu and X. Shen, *ACS Appl. Mater. Interfaces* **9**, 253 (2017).
27. M. Yu, S. Zhao, H. Feng, L. Hu, X. Zhang, Y. Zeng, Y. Tong and X. Lu, *ACS Energy Letters* **2**, 1862 (2017).
28. M. Chen, W. Zhou, M. Qi, J. Yin, X. Xia and Q. Chen, *J. Power Sources* **342**, 964 (2017).
29. M. Chen, D. Chao, J. Liu, J. Yan, B. Zhang, Y. Huang, J. Lin and Z. X. Shen, *Adv. Funct. Mater.* **27**, 1606232 (2017).
30. X. Xia, Y. Zhang, Z. Fan, D. Chao, Q. Xiong, J. Tu, H. Zhang and H. J. Fan, *Adv. Energy Mater.* **5**, 1401709 (2015).
31. X. Xia, Y. Zhang, D. Chao, Q. Xiong, Z. Fan, X. Tong, J. Tu, H. Zhang and H. J. Fan, *Energy Environ. Sci.* **8**, 1559 (2015).
32. X. H. Xia, D. L. Chao, Y. Q. Zhang, Z. X. Shen and H. J. Fan, *Nano Today* **9**, 785 (2014).

Chapter 6
Hard Carbon Anode Materials for Sodium-Ion Batteries

Ismaila El Moctar*, Qiao Ni*, Ying Bai*,
Feng Wu*, and Chuan Wu*,†

Recent results have shown that sodium-ion batteries complement lithium-ion batteries well because of the low cost and abundance of sodium resources. Hard carbon is believed to be the most promising anode material for sodium-ion batteries due to the expanded graphene interlayers, suitable working voltage and relatively low cost. However, the low initial coulombic efficiency and rate performance still remains challenging. The focus of this review is to give a summary of the recent progresses on hard carbon for sodium-ion batteries including the impact of the uniqueness of carbon precursors and strategies to improve the performance of hard carbon, and highlight the advantages and performances of the hard carbon. Additionally, the current problems of hard carbon for sodium-ion batteries and some challenges and perspectives on designing better hard-carbon anode materials are also provided.

Keywords: Sodium-ion battery; hard carbon; strategy; challenges and perspectives.

---

*School of Materials Science and Engineering, Beijing Institute of Technology, Beijing 100081, P. R. China.
† chuanwu@bit.edu.cn

1. Introduction

Today, our way of life has become completely dependent on energy and this quality of life has contributed to an explosion of global demography. It is important to develop renewable energies such as solar energy, tidal power and wind energy. Among these systems, electrochemical storage is an effective and practical solution.[1] Among the existing and most efficient electrochemical storage devices, secondary batteries technique is one of the most promising means for energy storage on a large-scale because of flexibility, high-energy conversion efficiency, and simple maintenance.[2]

In the past three decades, the rechargeable energy market has been dominated by lithium-ion batteries (LIBs).[3] Highenergy density, long cycle life and good electrochemical performance make LIBs the most coveted.[4] The suitable electrochemical potential makes it a congruous component of high-energy-density rechargeable LIBs.[5,6] However, the increasing cost and rarity of lithium resource restrain its development. Many efforts have been made to explore new alternative energy storage technologies.[7] Sodium-ion batteries (SIBs) are based on the same principle as LIBs because Na and Li lie in the same main group.[8] It is emerging worldwide again today and the numbers of publications have increased exponentially since 2010.[9] The relatively low cost and abundance of sodium resources make SIBs a good complementation to LIBs in the field of electrical energy storage.[10] However, the sodium ion has a larger ionic radius than that of the lithium. Therefore, the design of suitable host materials with larger space for intercalating and accommodating sodium ions is very difficult.[11]

Among many anode materials for SIBs, hard carbon is very attractive due to its lower redox potential, structure stability, the expanded graphene interlayers and relatively low cost.[12] However, the overall performance of SIBs is limited because of the poor rate performance of hard carbon anodes. Therefore, it is emergent to develop a high-performance hard carbon anode for SIBs.[13] Although various carbonaceous materials such as graphene, carbon nanotubes and carbon nanofibers have been widely investigated for SIBs as anode materials, natural biomass-derived carbons also show good electrochemical performance for SIBs.[14] Graphite carbon materials have been known as the promising electrode materials for LIBs, with respect to their unique electrochemistry functions and low cost.[15,16] Although in the rocking-chair model of LIBs, the graphitic anode possesses the chemical formula of LiC_6 after intercalation, Na ions cannot intercalate into the graphite owing to the absence of stable Na–C binary compounds.[17] The intercalation of sodium ions into graphite is not favored compared to lithium, and many studies have attributed this behavior to the bigger size of sodium element.[18] The discovery of anode materials with high theoretical capacity, environment friendliness and lowcost for SIBs is a major challenge.[19] Although, hard carbon, containing closed nanopores in itself has been exanimated as anode electrodes for SIBs, the relatively low initial coulombic efficiency (ICE <90%) and poor rate performance are less sufficient for practical applications.[20] Among different investigated carbon materials, hard-carbons are unquestionably the most promising candidates.[21] In order to reduce the cost and enhance the sustainability of these materials, it is known that using biomass as a source of hard carbon could be a very good alternative.[22,23]

The main scope of this review is to summarize the recent progress on hard carbon for SIBs including the impact of the uniqueness of carbon materials and the strategies to improve the carbon structure. The focus is placed on hard carbons especially hard carbons derived from biomass to highlight the advantages and the performances of the hard carbon as anode materials for SIBs.

2. Impact of Precursor's Uniqueness

In general, the aromatic hydrocarbons are typically forms of soft carbons with certain degrees of graphitization, such as coal tar, asphalt, while some polymers (including sp3 C) are often characterized by hard carbon behavior, such as cellulose, sugar polymers and vegetable biomass. Graphite is a kind of long-range order layered structure, while hard carbon is a kind of short-range order structure even in a limited condition.

In terms of capacity and reversibility, much progress has been made using hard carbon as anode materials for SIBs. Among various hard carbon materials proposed for anode application, biomass-derived hard carbon has received particular attention. Most of the materials used as precursors for hard carbons are either sugars, like sucrose and glucose; peels of fruits produced and consumed in certain parts of the world in large quantities (apples, bananas, shaddock peanuts etc.) or the most abundant natural polymers such as cellulose and lignin, derived from wood processing and other sources, as listed in Table 1.[24] Figure 1 summarizes the typical precursors-derived hard

Table 1. The typical precursors derived hard carbon for SIBs.

Classification	Precursors	Highest charge capacity (mAh g^{-1})	Capacity retention (%)	Current density (mA g^{-1})	References
Polymer	PAN	393	69.6 (after 100 cycles)	20	16
	Pitch	284	98 (after 100 cycles)	30	29
	PVC nanofiber	271	77.8 (after 120 cycles)	12	32
	PVP	393	98.2 (after 100 cycles)	20	63
	Polyaniline	275	77 (after 500 cycles)	50	81
	Sodium polyacrylate	341	68.3 (after 100 cycles)	50	82
	Tire	256	66 (after 100 cycles)	20	83
	P doped PAN	284	87.8 (after 200 cycles)	50	90
Natural biomass	Corn stalks	321	76.9 (after 100 cycles)	50	35
	Sorghum stalk	255	96 (after 50 cycles)	20	42
	Pinecone	370	96.9 (after 120 cycles)	30	43
	Mangosteen	330	98 (after 100 cycles)	20	44
	Argan shell	300	94.1 (after 70 cycles)	25	45
	Rice husk	276	93 (after 100 cycles)	25	62

Table 1. (*Continued*)

Classification	Precursors	Highest charge capacity (mAh g⁻¹)	Capacity retention (%)	Current density (mA g⁻¹)	References
	Dandelion	372	85.3 (after 300 cycles)	50	84
	Corn straw piths	310	88.4 (after 100 cycles)	50	85
	Ramie fiber	122	79 (after 100 cycles)	100	86
	Cotton	315	97 (after 100 cycles)	30	87
	Kelp	334	59 (after 300 cycles)	200	88
	Coconut endocarp	314	92% (after 200 cycles)	50	89
Sugar	Cellulose	300	84 (after 500 cycles)	37.2	23
	Sucrose	290	97 (after 150 cycles)	60	47
	D-glucose	330	53 (after 300 cycles)	25	50
	Sucrose/Go	289	95 (after 200 cycles)	20	52

carbon for SIBs, including polymersderived hard carbon, natural biomass-derived hard carbon, and sugar precursors-derived hard carbon.

2.1. *Polymer precursors derived hard carbon*

Polymers-derived hard carbon has been reported as electrode materials and showed great potential for SIBs. This high performance is owed to their uniform pore dimension and high surface area. The leonardite/humic acid (LHA)-derived hard-carbon has been reported by Zhu *et al.*[28] From the XPS spectra shown in Figs. 2(a) and 2(b), carbon and oxygen are the dominant elements in LHA. Three types of carbon species are shown in C1s spectrum: C–C (284.4 eV), C–O

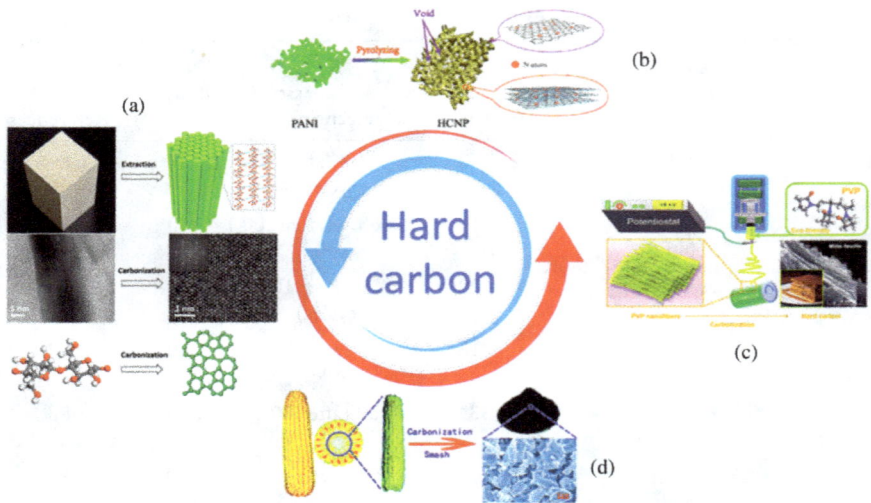

Fig. 1.   Schematic illusion of the typical precursors-derived hard carbon for SIBs: (a) sugar-derived hard carbon (b) precursors-derived hard carbon;[19] (c) polymers-derived hard carbon;[25,26] (d) biomass-derived hard carbon.[27] Reprinted with permission.

(285.1 eV) and O–C=O (288.7 eV), which demonstrates the surface chemical components of the hard carbon (Fig. 2(b)). Figure 2(c) shows the XRD patterns of the LHA at different pyrolysis temperatures. The (002) peak shifts to higher angle and becomes narrow with increasing temperature, indicating the structure arrangement from disordering to short-range ordering. Note that, with increasing temperature, the fraction of the total capacity contributed by the slope decreases (Fig. 2(d)). For another, a high specific capacity of 284 mAh g^{-1} with a capacity retention of 94% after 100 cycles and an ICE of 88% has been presented using hardcarbon from a mixture of pitch and phenolic resin by Li *et al.*[29] Hao and co-workers have demonstrated that the polyacrylonitrile/humic acid-derived carbon material used as anode material can exhibit a reversible capacity of 261 mAh g^{-1} at 0.02 Ag^{-1} with an ICE of 69.6% after 100 cycles as shown in Figs. 2(c) and 2(d).[30] As we know, reducing the carbon size is particularly important for improving electrochemical performance. For example, a hard carbon nanoparticle (HCNP) showed a high capacity at

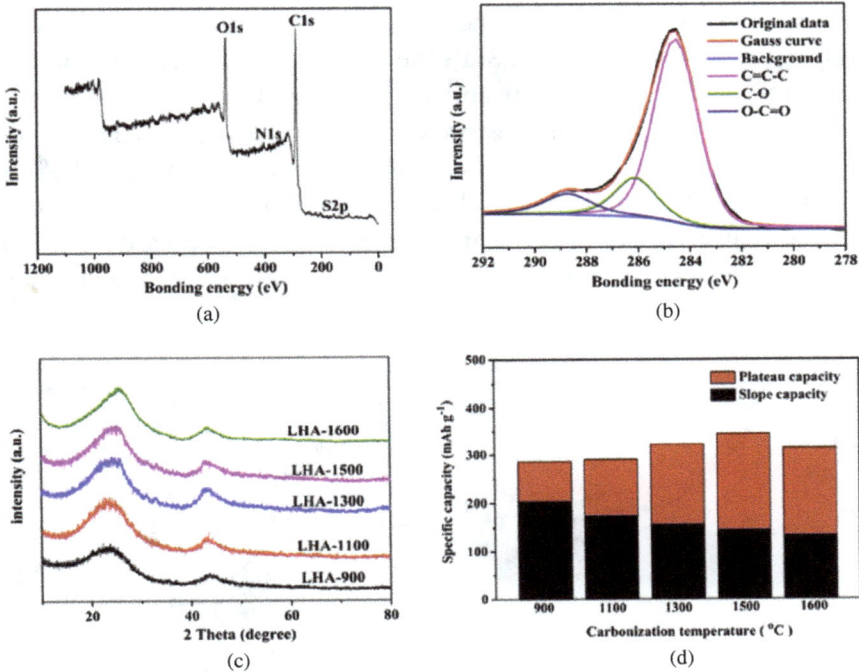

Fig. 2. (a) XPS curves spectra of LHA; (b) high-resolution XPS curves of the Cls for LHA; (c) XRD analysis of different LHA samples; (d) specific capacity from the plateau and slope contributions of LHA electrodes at 0.1 C.[28] Reprinted with permission.

voltages lower than 1.2 V.[31] Bai's group has prepared two kinds of hard carbon via carbonization process of commercialized polyvinyl-chloride (PVC) and PVC nanofiber from electrospinning. The obtained hard carbon from PVC nanofiber showed a reversible capacity of 271 mAh g^{-1} and ICE of 69.9%, while a reversible capacity of only 206 mAh g^{-1} and ICE of 60.9% were obtained from commercialized PVC.[32] Similarly, by electrospinning process, a hard carbon with unique morphology has been obtained from low-cost and environment friendly polyvinylpyrrolidone (PVP) nanofibers.[33] The optimized sample exhibits better particle size and low surface area, showing a high reversible capacity of 271 mAh g^{-1} with 94% capacity retention ratio over 100 cycles. In 2016, Simone and his co-workers reported a hard carbon prepared via pyrolysis of an abundant organic polymer.[23] As showed in Fig. 3(a), the scattering intensity is a function of electron density variation and increases with increasing pyrolysis temperature. All hard carbons present the similar first discharge specific capacity around 360 mAh/g. The highest ICE value of 84% is obtained for the sample of HC-1600 (Fig. 3(b)). Figure 3(c) demonstrates that the specific capacity increases when the carbonization temperature increases. Meanwhile, the micropore size increases while

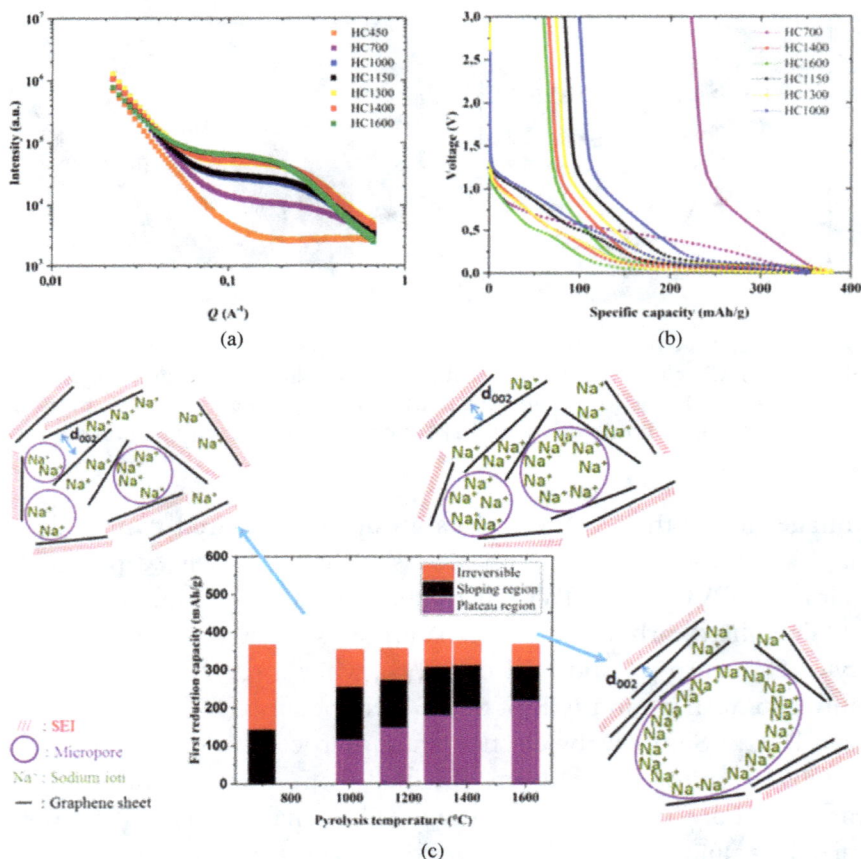

Fig. 3. (a) Intensity per sample mass as a function of the scattering vector Q for all hard carbons; (b) First charge–discharge curve of hard carbons cycled at a C/10 rate; (c) description of pyrolysis temperature impact on structure and electrochemical performance of hard carbon.[23] Reprinted with permission.

the graphene planes become closer, which leads to the reduction of the slope capacity and an increase of the low voltage plateau capacity with increasing temperature. It can be seen that, through these few examples, the characteristics of carbon, including the degree of graphitization, the structural or textural disorder and porosity have a great influence on the capacity and mechanism of sodium insertion.

2.2. *Natural biomass-derived hard carbon*

As one of the representatives of carbon-based materials, natural biomass has attracted significant attention. Most of the recent reports about hard carbon proposed natural biomass as good strategy for an excellent sodium-ion storage. Treated under different carbonization temperatures, the biomass precursors exhibited different morphologies. Most of the research papers reported that the specific capacity of biomass-derived hard carbon is related to the carbonization temperature.[34] Qin *et al.* synthesized analogous graphite carbon sheets from stalks as the high-performance SIBs anodes. The sample treated at 1200°C showed a stable reversible capacity of 231 mAh g^{-1} after 200 cycles at 50 mA g^{-1} with good rate capability and excellent long-term cycling stability.[35] A hard carbon derived from pitch has been prepared by Yang and his co-workers and used as electrode materials in SIBs. The electrochemical tests showed a high initial discharge capacity of 268.4 mAh g^{-1} with high initial efficiency coulombic of 79.2% at a current density of 20 mA g^{-1}.[16] Carbon nanosheet frameworks can synergistically enhance the reversible capacity and cycle stability of SIBs. For example, Hu and co-workers prepared the carbon nanosheet frameworks derived from sodium alginate.[36] The materials obtained as carbon showed a high reversible capacity of 216 mAh g^{-1} at 100 mA g^{-1}; a good cycle stability of 160 mAh g^{-1} at 100 mA g^{-1} after 340 cycles and excellent rate capability of 66 mAh g^{-1} at 5000 mA g^{-1}. However, hard carbon prepared from ramie fibers and corn cobs by carbonization in Ar atmosphere at 700°C and used as anodes materials for SIBs showed the low reversible capacity of, respectively, 122 mAh g^{-1} and 193 mAh g^{-1} with a similar cycling stability.[37] From above report, we can see that the carbonization temperature is an important factor for hard carbon synthesis.

Li and co-workers study showed that the hard carbon synthesized via simple calcination process of natural cotton can deliver a reversible capacity of 315 mAh g^{-1} with good rate capability. This excellent electrochemical performance can be attributed to the unique tubular structure.[38] According to recent works of Wang and co-workers the turbostratic lattice of kelp-derived hard carbon with larger interlayer distance (3.9–4.3 Å) delivered a high reversible capacity; a stable capacity of 96 mAh g^{-1} at 1000 mA g^{-1} and good cycling performance of 205 mAh g^{-1} after 300 cycles at 200 mAh g^{-1}.[39] Zhu et al. synthesized a high-performance hard carbon from waste apricot shell. As an anode for SIBs, it exhibits a promising anode performance with a superhigh reversible capacity of 400 mAh g^{-1}. What's more, it shows an ICE of 79% and an excellent cycling stability.[40] Zheng's group fabricated a hard carbon via simple pyrolysis of macadamia shell and tested in the full cell as an electrode material for sodium-ion batteries. The electrode showed a capacity of 314 mAh g^{-1} with an ICE of 91.4%.[41] Recently, Zhu et al. reported a good rate capability of 172 mAh g^{-1} at 200 mA g^{-1} with good cycling performance of 245 mAh g^{-1} after 50 cycles using the sorghum stalk-derived hard carbon as anode material.[42] Very recently, Zhang's group have synthesized hard carbon derived from pinecone via a simple carbonization process, the sample treated at 1400°C showed a reversible capacity of 370 mAh g^{-1} at a current density of 30 mA g^{-1} with good ICE 85.4%.[43]

Recently, Wang et al. synthesized a hard carbon by simple pyrolysis of biowaste mangosteen shell. The obtained hard carbon showed a reversible capacity of 330 mAh g^{-1} at a current density of 20 mA g^{-1}; a capacity retention of 98% after 100 cycles.[44] Mouad's group prepared a hard carbon through carbonization and Hydrochloric acid (HCl) treatment of argan shell which showed a reversible capacity of 300 mAh g^{-1} at 25 mA g^{-1} and capacity retention of 94.1% after 70 cycles.[45] From above examples, we can see that the hard carbon derived from biomass is promising anode material for SIBs. This advantage is related to its contents of cellulose and hemicellulose which can be converted into carbon materials via carbonization process.

2.3. *Sugar precursors derived hard carbon*

Among many carbonaceous materials, the hard carbon-derived sugar precursors showed high environmental impacts.[46] The carbonization temperature plays an important role in the specific capacity of hard carbon. For example, in 2014 Li and co-workers investigated the influence of the carbonization temperature of the microstructure and electrochemical performance of the nano-dispersed hard carbon derived from sucrose. The hard carbon showed a high reversible capacity of 220 mAh g^{-1} and an excellent cycling performance with a capacity retention of 93% after 100 cycles.[47] In 2015, Ponrouch's group reported that hard carbon prepared from sugar via pyrolysis, which showed a high reversible capacity of 450 mAh g^{-1} at 20 mA g^{-1} and a rate capability of 340 mAh g^{-1} as anode material in SIBs. The cycling stability is also excellent.[48] Prabakar and his co-workers investigated the porosity and morphology of sucrose-based hard carbon (SHC). The hard carbon with good cycling performance as described in Figs. 4(a) and 4(b) was prepared from bicarbonate and table sugar.[49] Recently, the performance of Dglucose-derived hard carbon was evaluated in LIBs and SIBs for comparing. The best performance has been obtained for SIBs using NaPF$_6$ as an electrolyte which is attributed to its appropriate morphology and cycling performance as shown in Figs. 4(c) and 4(d).[50] Palacín' group prepared hard carbon from sugar pyrolysis, which exhibits a specific capacity over 300 mAh g^{-1} at 0.1 C after 300 cycles.[51]

Recent studies showed that reducing the surface area of hard carbon at a relatively low pyrolysis temperature is very critical. Luo's group demonstrated an effective strategy to reduce the surface area of sucrose-derived hard-carbon by introducing graphene oxide (GO) suspension into the sucrose solution.[52] Zhang *et al.* reported a hard carbon from bacterial cellulose (BC) and 2, 2, 6, 6, -tetra-methylpiperidine-1-oxyl (TEMPO) oxidized-BC via a simple carbonization process. The results reveal that the introduction of carbonyl is an effective strategy to enhance the electrochemical performance of hard carbon.[53] Recently, Zhu and co-workers transformed the ordered cellulose nanocrystal (CNCs) into porous

(a)

(b)

(c)

(d)

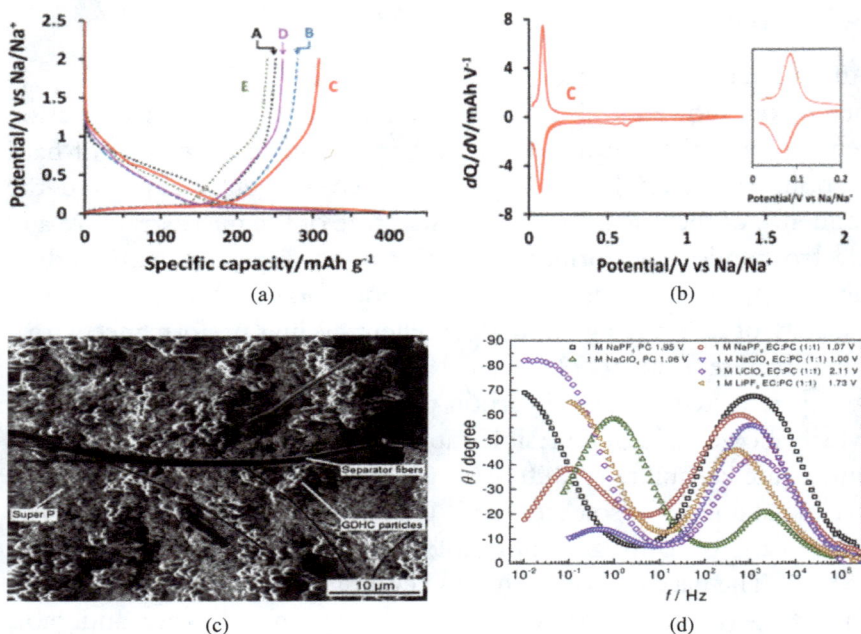

Fig. 4. (a) C/D curses of (A) SHC, (B) NSHC-2.5, (C) NSH-5, (D) NSHC-10 and (E) NSHC-20 at 20 mA/g; (b) dQ/dV curves for 1st 2nd 5th C/D curves of (C) NSHC-5;[49] (c) SEM image of electrochemically Na-charged GDHC electrode; (d) Blode phase angle plots of the studied half cells measured at open circuit potential.[50] Reprinted with permission.

carbon at a relatively low carbonization temperature of 1000°C used as the anode for SIBs. The CNCs-derived porous carbon showed the highest reversible capacity (340 mAh g^{-1} at 100 mA g^{-1}). The rate capability and cycling stability are also excellent. The excellent electrochemical performance is ascribed to the larger interlayer spacing and porous structure.[19]

3. Strategies for Enhancing the Electrochemical Performance of Hard Carbon

3.1. *Hetero-atom doped hard carbon*

Presently, a lot of efforts have been made on improving hard carbon's sodium storage capacity, where the impacts of the uniqueness of carbon precursors, morphologies and heteroatom doping have been highlighted.[54] The surface modification can be considered as an effective strategy to improve the electrochemical performance of hard carbon. For improving the energy storage of SIBs, the hierarchical porous nitrogen-rich carbon nanospheres from acidic etching of metal carbide and carbon hybrid nanoarchitecture were investigated.[55] Ou *et al.* applied a nitrogen-doped porous carbon (NDPC) from the horn. The obtained hard carbon offered good cycling and rate performance which is attributed to its high charge transfer resistance described in Fig. 5(a), and large surface area by nitrogen doping as shown in the Fig. 5(b).[56] The hierarchical porous nitrogen-doped carbon from horn comb has been investigated as an anode in SIBs, which exhibited a reversible capacity of 400 mAh g^{-1}.[57] The high capacity obtained from above (NDPC) and hierarchical nitrogen-doped porous carbon is owed to presence of different oxygen functional groups. The oxygen functional groups improve the sodium storage capacity by participating in the surface redox reaction with sodium ions. In another work, Wang and co-workers reported a case study of the nitrogen and oxygen functional groups of high surface area carbon nanosheets. The results showed that the surface area of functionalized carbon nanosheets is significantly reduced to 840 m2 g^{-1}. However, the electrodes deliver a capacity of 162 mAh g^{-1} at 1 Ag^{-1} after 2000 cycles and retain a capacity of 49 mAh g^{-1} at 10 Ag^{-1}.[17] Recently, Hu's group improved the performance of the polyurethane foam derived hard carbon. They used NaCl for hard carbon intercalation via a simple synthesis process as shown in Fig. 5(c).[58] Zhang's group fabricated a nitrogen porous doped carbon using covalent-organic frameworks as precursors and used as the anode in SIBs.[59] The electrode material presents different semicircle corresponding to the solid electrolyte interface (SEI) layers resistance (R_f) and large semicircle corresponding to the charge transfer resistance (R_{ct}) as described in the Fig. 5(d). In summary, introducing nitrogen in carbon can dramatically promote capacity storage on the surface due to the increased active sites produced by *N*-doping. In addition, after *N*-doping, the electronegativity has developed than undoped counterparts, which form a stronger interaction between ions.[60]

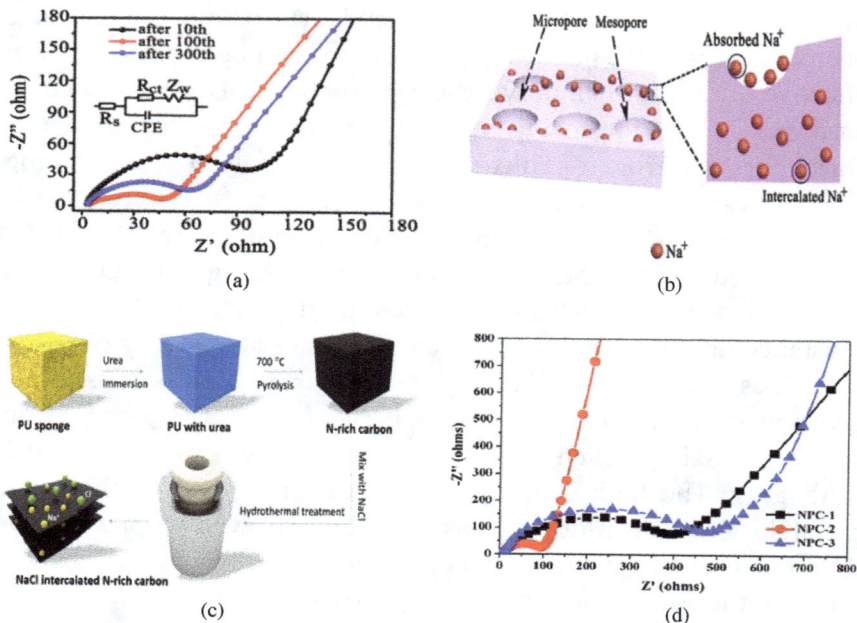

Fig. 5. (a) Electrochemical impedance spectra of NDPC after various cycles; (b) schematic illustration of Na+ transportation and storage in the synthesized NDPC;[56] (c) schematic illustration of the NaCl intercalation nitrogen-rich hard carbon synthesis;[58] (d) Nyquist plots of NPCs electrode for SIBs after 100 cycles.[59] Reprinted with permission.

Ruan's group reported a reversible capacity of 520.1 mAh g^{-1} for SIBs using a nitrogen–sulfur dual doped carbon (NS-C).[61] The excellent capacity and cycling stability is ascribed to its appropriate structure caused by nitrogen and sulfur doping. Guo *et al.* reported nitrogen-doped porous carbon from mineral-rich egg yolks. The prepared hard carbon as anode material in SIBs showed an initial reversible capacity of 208 mAh g^{-1}.[17] The obtained high capacity is owed to the electrical conductivity enhanced by heteroatoms incorporation in the carbon.

Li *et al.* reported that doping PO$_x$ into the local structure of hard carbon upgrades its reversible capacity from 283 to 359 mAh g^{-1}. Through XANES and neutron total scattering, it confirmed that the doped PO$_x$ is redox inactive, thus not contributing to the higher capacity, but doping can lead to turbostratic nanodomains more defective and increased (002) interlamellar spacing.[62] Very recently, Wu's group reported that P=O and P–C bonds can form in graphitic layer when doped with phosphorus, which is fabricated by electro-spinning method.[63] The phosphorus functionalized hard carbon can deliver a high reversible capacity of 393.4 mAh g^{-1} as anode for SIBs is accompanied by the increased electron density around the Fermi level.

3.2. *Activated hard carbon*

Among many methods, activation is the most convenient for the surface modification of carbon materials.[64,65] In order to improve the performance of biomass waste-derived carbon for energy storage, different activation methods such as chemical activation and physical activation have been utilized.

3.2.1. *Chemical activation*

Chemical activation is the most used method, which usually includes three stages: firstly, the raw materials are carbonized; secondly, the obtained carbon material is mixed with a chemical activating reagent such as KOH, H_3PO_4, $ZnCl_2$ or $KHCO_3$; finally, the heating is done in an inert gas at a high temperature.[20] The impregnation of precursors in chemical reagent enhances the surface area.[66] Since 1997, Tsai and coworkers used the chemical activation to synthesis activated carbons from waste corncobs using $ZnCl_2$ as a chemical agent. The obtained carbon material was essentially microporous carbon material.[67] It is important to know that the purity will improve as the pyrolysis temperature increased since some functional group is eliminated. By simple pyrolysis of H_3PO_4 treated biomass pomelo peels, a porous carbon material has been synthesized by Kun-Lei *et al.* The obtained hard carbon used as the anode for SIBs showed a good cycling stability and rate capability, delivering a capacity of 181 mAh g^{-1} at 200 mA g^{-1} after 220 cycles and retaining a capacity of 71 mAh g^{-1} at 5 A g^{-1}.[12] Wang and co-workers reported a hard-carbon from biomass peanut skin via hydrothermal and chemical activation method as described in Fig. 6(a). The KOH-activated peanut skin-derived carbon can exhibit a good rate capability and cycling stability: a high initial charge capacity of 431 mAh g^{-1} at 0.1 Ag^{-1}, retaining a reversible capacity of 47 mAh g^{-1} at 10 Ag^{-1}, and showing a capacity retention of 83–86% after 200 cycles which attributed to its disordered structure described in Fig. 6(b).[17] Lv *et al.* prepared a hard-carbon via similar method using peanut shells. A porous hard carbon as shown in Figs. 6(c) and 6(d) has been obtained.[68] Recently, Xiang and coworkers have prepared the KOH activated orange peel-derived hard carbon. Used as anode material for SIBs, the results showed an initial capacity of 497 mAh g^{-1} at 0.5 Ag^{-1} and good cycling stability (1000 at 1 Ag^{-1}). The excellent performance can be attributed to the disorder activated hard carbon structure.[69] In 2016, Zheng and co-workers synthesized a filter paper-derived micro nanostructure hard carbon by simple carbonization process followed by KOH treatment. The

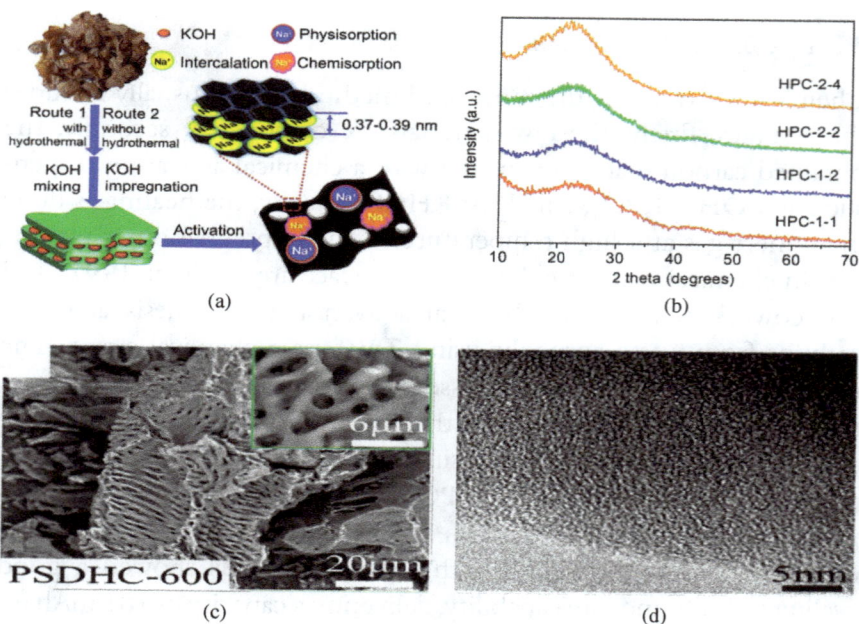

Fig. 6. (a) Schematic illustration of the formation of hierarchical porous carbon and the relevant charge storage mechanism; (b) XRD patterns of the HPC specimens;[17] (c) SEM image of SDHC-600; (d) HRTEM image of SDHC-600.[68] Reprinted with permission.

obtained carbon material used as the anode in sodium ion half-cell exhibited a first-cycle Coulombic efficiency as high as 80% and a reversible capacity of 268 mAh g^{-1} after 100 cycles at 20 mA g^{-1}. The performance can be attributed to the unique micro-nano structure by high temperature pyrolysis and KOH activation treatment.[70]

Physical activation includes the carbonization at low temperature under an inert atmosphere to remove non-carbon elements and the activation at high temperature by employing CO_2 or steam as activating reagent.[20] In 2014, Bommier and co-workers synthesized hard carbon via simple dehydrating sucrose for 24 h under atmospheric conditions at 180°C. Furthermore, through CO_2 activation experiments, it confirmed that the porosity is in close contact with reversible capacity, and the Na-ion storage depends on the absence of pores detectable in sucrose-derived hard carbon.[71] The physical activation method is environment friendly. However, the obtained surface area on the carbon is not enough compared with chemical activation, in general.

3.3. *Impact of electrolytes on hard-carbon*

Sodium-based electrolytes were investigated in SIBs using hard-carbon as an anode. Although aqueous electrolytes have little risk of ignition or explosion of the battery during a possible heating of the device,[72] organic electrolytes have good ionic conductivity and a wide range of potential stability.[73] This is the reason why organic electrolytes are used in most studies on SIBs. The most common electrolyte solvents are propylene carbonate (PC), ethylene carbonate (EC) and dimethyl carbonate (DMC). Various additives such as fluoroethylene carbonate (FEC), difluoroetyhene carbonate (DFEC) and vinylene carbonate (VC) are well-known to be the efficient electrolyte additives. However, it has been proved that FEC is the only efficient electrolyte additive among above three additives due to improved passivation and suppression of side reactions.[74] As shown in Fig. 7(a), the capacity and cyclic stability are remarkably improved when 2 vol.% FEC was added into PC. Komaba and co-workers studied the hard carbon in the $NaClO_4$ as electrolyte salts, and the electrochemical properties of the hard carbon in EC, PC, Butyl Carbonate

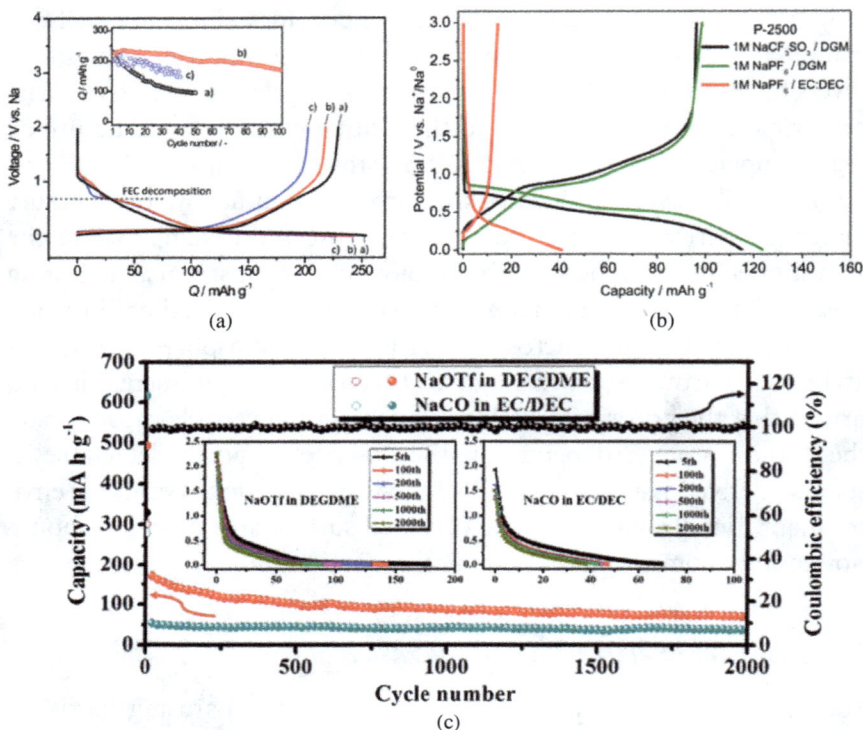

Fig. 7. (a) Initial charge–discharge curves of hard-carbon in 1 M NaClO$_4$ PC solution without (a), with 2% (b), and 10 vol.% FEC (c), the inset presents the cyclic stability of the three electrodes for 100 cycles;[74] (b) Charge–discharge curves of P-2500 hard carbon in electrolytes of 1 M NaCF$_3$SO$_3$ in DGM, NaPF$_6$ in DGM and NaPF$_6$ in EC: DMC (1:1) at rate of 1C;[77] (c) Long-term cycling performance of the electrodes for 2000 cycles at rate of 7 C in different electrolytes. The insets are discharge curves at different cycles in the ether- and ester-based electrolytes.[31] Reprinted with permission.

(BC) based electrolyte and EC: DMC (1:1), EC: EMC (1:1), EC: DEC (1:1) mixed electrolyte. It was found that the hard carbon anode exhibited excellent reversible capacity and capacity retention in PC, EC and EC: DEC-based electrolyte at room temperature. In low temperature, the electrolyte with PC as solvent is more suitable for hard carbon.[72]

Recently, a lot of work has been reported to improve Na$^+$ storage in natural graphite by substituting the ester-based electrolytes with ether-based electrolytes.[75,76] As it demonstrated that Na^+ can co-intercalate into graphite layers with solvent. Recently, Cabello and his co-workers reported a hard carbon from trash coke. Their investigation showed that the hard-carbon with $NaPF_6$ as an electrolyte salt was promising power source because of its appropriate morphology and cycling performance as shown in Fig. 7(b).[77] Zhu's group reported that the capacity loss for sodium storage in hard carbon in ester-based electrolytes can be lowered in etherbased electrolytes.[31] Figure 7(c) exhibits the cycle stability of the hard carbon at high rate of 7 C in the ether- and esterbased electrolytes. It is obviously observed that the capacity is about twice in the ether-based electrolyte than that of esterbased electrolyte in the 2000 cycles.

The properties of hard carbon (HC) for sodium secondary batteries using $NaFSA-C_1C_3pyrFSA$ (FSA: bis (fluorosulfonyl) amide, C_1C_3pyr: N-methyl-N-propylpyrrolidinium) ionic liquids were also studied. The results showed that a reversible capacity of 260 mAh g^{-1} at a constant rate of 50 mA g^{-1} can be obtained, maintaining approximately 95.5% of the initial capacity after 50 cycles.[78] The ionic liquid with an excellent thermal stability and negligible volatility are promising electrolytes for SIBs.[79,80]

4. Conclusion and Personal Outlook

Nowadays, carbon materials are the most commercially promising anode materials for SIBs. The microstructure of hard carbon has crucial influence on electrochemical behavior and capacitance effect. As is well-known, hard carbon can be considered as a sponge with closed pore, whose matrix is formed by a mutually connected nano-carbon crystal, for lithium- and sodium-ion battery, in the high and low potential regions. However, most of the surface of carbon microcrystalline is electrolyte inaccessible, and a solid electrolyte interphase (SEI) may be generated between the surface of material and the electrolyte, which would hamper the ion and electron transport. Therefore, hard carbon materials with low specific surface area are often suitable for battery systems. In contrast, supercapacitors usually require hard carbon materials with a large specific surface area and porosity due to the rapid kinetics.[91]

In this review, we mainly focused on the current developments and modified strategies of hard carbon. Although some achievements have been obtained, in an effort to practical utilization of hard anodes for SIBs, the following challenges remains: (1) There is still considerable controversy over the mechanism of sodium storage of hard carbon materials, and it may be necessary to conduct in-depth and direct analysis of structures and images in order to provide more conclusive evidence; (2) The relatively low ICE and poor rate performance still restrict its commercialization; (3) The research work on electrolyte and additive of SIBs is not enough. To improve the ICE, the design of hard carbon with low surface area while containing rich reversible sodium storage active sites is especially crucial, which can slow down the side effects. For another, through hetero-atom doping has been proved to be another effective method to stabilize the carbon structure and increase layer spacing. Physical and chemical activation are high-efficient ways for the surface modification of hard carbon, especially for biomass-derived hard carbon. By optimizing carbonate-based electrolyte with new additive or substituting the ester-based electrolyte with ether-based electrolyte can possibly tailor the formation of SEI film and improve Na^+ storage.

Finally, the eco-friendly biomass-derived hard carbon materials should receive wide attention. It is believed that synthesizing the hard carbon from abandoned biomass precursors with efficient preparation method would be a promising approach for SIBs application.

Acknowledgment

This work was supported by the National Basic Research Program of China (Grant No. 2015CB 251100).

References

1. L. Cao *et al.*, *J. Alloys Compd.* **695**, 632 (2017).
2. J. Tang *et al.*, *Curr. Opin. Chem. Eng.* **9**, 34 (2015).
3. H. Pan *et al.*, *Energy Environ. Sci.* **6**, 2338 (2013).
4. Y. Yao *et al.*, *Nano Energy* **17**, 91 (2015).

5. J.-C. Li *et al.*, *Func. Mater. Lett.* **10**, 1750054 (2017).
6. A. Hu *et al.*, *J. Energy Chem.* **27**, 203 (2018).
7. T. G. Goonan and U. S. Geol, *Survey* **1** (2013).
8. Q. Ni *et al.*, *Adv. Sci.* **4**, 1600275 (2017).
9. A. Eguia-Barrio *et al.*, *MRS Adv.* **2**, 1165 (2017).
10. T. Vogl *et al.*, *J. Mater. Chem. A* **4**, 10472 (2016).
11. D. Kundu *et al.*, *Angew. Chem. Int. Ed.* **54**, 3431 (2015).
12. H. Shi *et al.*, *Funct. Mater. Lett.* **10**, 1750076 (2017)
13. Y. Li *et al.*, *Energy Storage Mater.* **7**, 130 (2017).
14. X. Zhou *et al.*, *J. Phys. Chem. C* **118**, 22426 (2014).
15. J. Song *et al.*, *Nano Energy* **40**, 504 (2017).
16. A. Ramos *et al.*, *Electrochim. Acta* **178**, 392 (2015).
17. H. Wang *et al.*, *Electrochim. Acta* **188**, 103 (2016).
18. S. Wenzel *et al.*, *Energy Environ. Sci.* **4**, 3342 (2011).
19. H. Zhu *et al.*, *Nano Energy* **33**, 37 (2017).
20. W. Tang *et al.*, *Mater. Res. Bull.* **88**, 234 (2017).
21. X. Zhou *et al.*, *Green Chem.* **18**, 2078 (2016).
22. S. Yang *et al.*, *Chem. Eng. J.* **309**, 674 (2017).
23. V. Simone *et al.*, *J. Energy Chem.* **25**, 761 (2016).
24. W. Li *et al.*, *Mater. Lett.* **65**, 3368 (2011).
25. L. Xiao *et al.*, *Nano Energy* **19**, 279 (2016).
26. Y. Bai *et al.*, *RSC Adv.* **7**, 5519 (2017).
27. P. Liu *et al.*, *J. Mater. Chem. A* **4**, 13046 (2016).
28. Y. Zhu *et al.*, *Carbon* **123**, 727 (2017).
29. Y. Li *et al.*, *Energy Storage Mater.* **2**, 139 (2016).
30. P. Y. Zhao *et al.*, *Electrochim. Acta* **232**, 348 (2017).
31. Y. E. Zhu *et al.*, *J. Mater. Chem. A* **5**, 9528 (2017).
32. Y. Bai *et al.*, *ACS Appl. Mater. Interfaces* **7**, 5598 (2015).
33. Y. Bai *et al.*, *RSC Adv.* **7**, 5519 (2017).
34. A. Ramos *et al.*, *Electrochim. Acta* **187**, 496 (2016).
35. D. Qin *et al.*, *RSC Adv.* **6**, 106218 (2016).
36. Y. Sun *et al.*, *Nat. Commun.* **4**, 1870 (2013).
37. Y. Gao *et al.*, *Carbon* **51**, 52 (2013).
38. X. Zhu *et al.*, *JOM* **68**, 2579 (2016).
39. S. Dutta *et al.*, *Energy Environ. Sci.* **7**, 3574 (2014).
40. Y. Zhu *et al.*, *Carbon* **129**, 695 (2018).
41. Y. Zheng *et al.*, *Nano Energy* **39**, 489 (2017).
42. X. Zhu *et al.*, *Green Energy Environ.* **2**, 310 (2017).
43. T. Zhang *et al.*, *RSC Adv.* **7**, 41504 (2017).

44. K. Wang *et al.*, *ACS Omega* **2**, 1687 (2017).
45. M. Dahbi *et al.*, *Electrochem. Commun.* **44**, 66 (2014).
46. J. F. Peters *et al.*, *Energy Environ. Sci.* **9**, 1744 (2016).
47. Y. Li *et al.*, *J. Mater. Chem. A* **3**, 71 (2014).
48. A. Ponrouch *et al.*, *Electrochem. Commun.* **54**, 51 (2015).
49. S. J. R. Prabakar *et al.*, *Electrochim. Acta* **161**, 23 (2015).
50. R. Väli *et al.*, *Electrochim. Acta* **253**, 536 (2017).
51. A. Ponrouch *et al.*, *Electrochem. Commun.* **27**, 85 (2013).
52. W. Luo *et al.*, *ACS Appl. Mater. Interfaces* **7**, 2626 (2015).
53. T. Zhang *et al.*, *RSC Adv.* **7**, 50336 (2017).
54. Z. Li *et al.*, *Chem. Commun.* **53**, 2610 (2017).
55. L. Ma *et al.*, *Nanoscale* **8**, 17911 (2016).
56. J. Ou *et al.*, *Micropor. Mesopor. Mater.* **237**, 23 (2017).
57. J. Ou *et al.*, *Electro. Mater. Lett.* **13**, 1 (2016).
58. M. Hu *et al.*, *Carbon*, **122**, 680 (2017).
59. X. Zhang *et al.*, *Carbon* **116**, 686 (2017).
60. Y. Xu *et al.*, *Nat. Commun.* **9**, 1720 (2018).
61. X. Deng *et al.*, *Carbon* **107**, 67 (2016).
62. M. K. Rybarczyk *et al.*, *J. Energy Chem.*, https://doi.org/10.1016/j.jechem.2018.01.025 (2018).
63. Y. Li *et al.*, *Adv. Energy Mater.* **8**, 1702781 (2018).
64. J. Wang *et al.*, *J. Mater. Chem. A* **5**, 2411 (2016).
65. J. A. Maciá-Agulló *et al.*, *Carbon* **42**, 1367 (2004).
66. E. Schroder *et al.*, Progress in biomass and bioenergy production (InTechOpen, 2011).
67. W. T. Tsai *et al.*, *Bioresource. Technol.* **64**, 211 (1998).
68. W. Lv *et al.*, *Electrochim. Acta* **176**, 533 (2015).
69. J. Xiang *et al.*, *J. Alloy Compd.* **701**, 870 (2017).
70. P. Zheng *et al.*, *Sci. Rep.* **6**, 35620 (2016).
71. C. Bommier *et al.*, *Carbon* **76**, 165 (2014).
72. S. Komaba *et al.*, *Adv. Funct. Mater.* **21**, 3859 (2011).
73. L. Lin *et al.*, *Func. Mater. Lett.* **10**, 1750052 (2017).
74. S. Komaba *et al.*, *ACS Appl. Mater. Interfaces* **3**, 4165 (2011).
75. H. Kim *et al.*, *Adv. Funct. Mater.* **25**, 534 (2015).
76. B. Jache *et al.*, *Angew. Chem. Int. Ed.* **53**, 10169 (2014).
77. M. Cabello *et al.*, *J. Power Sources* **347**, 127 (2017).
78. A. Fukunaga *et al.*, *J. Power Sources* **246**, 387 (2014).
79. C. H. Wang *et al.*, *Chem. Commun.* **52**, 10890 (2016).
80. S. T. Senthilkumar *et al.*, *J. Power Sources* **341**, 404 (2017).

81. L. Xiao *et al.*, *Nano Energy* **19**, 279 (2016).
82. Z. Yuan *et al.*, *J. Mater. Chem. A* **3**, 23403 (2015).
83. Y. Li *et al.*, *J. Power Sources* **316**, 232 (2016).
84. C. Wang *et al.*, *J. Power Sources* **358**, 85 (2017).
85. Y. E. Zhu *et al.*, *Ionics* **24**, 1075 (2018).
86. Q. Jiang *et al.*, *Appl. Surf. Sci.* **379**, 73 (2016).
87. Y. Li *et al.*, *Adv. Energy Mater.* **6**, 1600659 (2016).
88. P. Wang *et al.*, *J. Mater. Chem. A* **5**, 5761 (2017).
89. F. Wu *et al.*, *ACS Appl. Mater. Interfaces*, doi: 10.1021/acsami. 8b08380 (2018).
90. F. Wu *et al.*, *ACS Appl. Mater. Interfaces* **10**, 21335 (2018).
91. J. Ni *et al.*, *J. Power Sources* **223**, 306 (2013).

Chapter 7

Some MoS$_2$-Based Materials for Sodium-Ion Battery

Qing Li*, Xiaotian Guo*, Mingbo Zheng*,†, and Huan Pang*,‡

Sodium-ion batteries (SIB) play a promising role in the area of energy storage device researching. MoS$_2$-based materials are considered as one of the most attractive materials for high-performance SIBs owing to their high capacity, high cycle stability and excellent Coulomb effect. This review has summarized the synthesis of MoS$_2$-based materials (MoS$_2$, MoS$_2$/carbon-based materials, MoS$_2$/metal oxides, MoS$_2$/metal sulfides) to emphasize their electrochemical behaviors in SIBs.

Keywords: MoS$_2$-based materials; electrochemical performance; sodium-ion batteries.

1. Introduction

Owing to the ever-increasing depletion of energy resources, the sustainable and clean energy related technologies are being developed at an incredible rate by various research institutions.[1–5] For example, considerable efforts have been made to develop sodium-ion batteries (SIBs).[6–8] Molybdenum disulfide (MoS$_2$) and the composites have

*Guangling College, School of Chemistry and Chemical Engineering, Yangzhou University, Yangzhou, Jiangsu 225009, P. R. China.
†mbzheng@yzu.edu.cn
‡huanpangchem@hotmail.com

received extensive attention as an electrode material due to their high capacity, high cycle stability and excellent Coulomb effect.[3,9,10]

SIB, based on abundant reserves of sodium resources and price advantages, are gradually becoming excellent candidate energy storage devices for replacing lithium-ion batteries (LIBs) in recent years. However, there are still many obstacles that need to be overcome to reduce the gap between LIBs in the development of SIBs, especially in finding suitable electrode materials.[11–13] Interestingly, MoS_2 is a specific layered transition-metal dichalcogenide. Specifically, the hexagonal layer of Mo is sandwiched between two S layers to form this style structure. Based on this feature, it has become a hot topic in the field of SIBs as a functional material. However, with regards to its physical properties, MoS_2 is a non-volatile, insoluble compound. Obstacles have been formed in preparing the film, which requires finding a fast and effective method to solve this problem.[14–17]

In recent years, MoS_2 and its composites have made many encouraging breakthroughs in the field of SIBs. Herein, we review the typical synthetic strategies and their specific electrochemical performance for SIBs as electrode materials, from the view of structural design and electrochemical properties control. It is hoped that this review can provide an effective reference for researchers to study the typical MoS_2-based materials for SIBs. We summarize some recent advances of the MoS_2-based materials for sodium-ion battery, as shown in Table 1.

Table 1. Recent advances in sodium storage performance for MoS_2-based materials.

Materials	Morphologies	Cycling performance		Rate performance		
		Capacity (mAh g⁻¹)/cycles	Current (A g⁻¹)	Capacity (mAh g⁻¹)	Current (A g⁻¹)	Ref.
MoS_2	worm-like structure	410.5/80	0.0617	/	/	15
	micro structure	136/70	1	420	0.05	18
	nanoflowers	/	/	300	1	19
	microflowers	595/50	0.067	236	0.067	20
MoS_2/C	nanospheres	400/300	0.67	520	0.067	21

Table 1. (*Continued*)

Materials	Morphologies	Cycling performance Capacity (mAh g^1)/cycles	Current (A g^{-1})	Rate performance Capacity (mAh g^{-1})	Current (A g^{-1})	Ref.
	hollow-nanospheres	410/1000	4	/	/	22
	nanotubes	475/200	0.5	432	1	23
	multiple-nanospheres	322/600	1.5	573	0.2	24
MoS$_2$/G	sandwich-like structure	313/200	0.05	175	2	25
	nanoribbons	158/1500	5	372	0.5	26
MoS$_2$/HfO$_2$	nanosheet	636/50	0.1	/	/	27
MoS$_2$/TiO$_2$	nanowire	191/100	0.02	/	/	28
MoS$_2$/TiO$_2$/CC	flower-like nanosheets	1392.8/150	0.2	392	0.8	29
MoS$_2$/SnO$_2$/C	sandwich-like nanosheets	230/450	1	/	/	30
MoS$_2$/Fe$_3$O$_4$/G	nanosheet	388/300	0.1	468	0.1	31
Fe2O$_3$/C/MoS$_2$/C	hydrangea-like nanoflakes	498/200	0.2	/	/	32
MoS$_2$/MoO$_3$/NC	amorphous	538.7/200	0.3	/	/	33
SnS/MoS$_2$	yolk-shell structure	396/100	0.5	476	0.2	34
Ni$_3$S$_2$/MoS$_2$	heterostructure	207/400	5	283	5	35
MoS$_2$/Ni$_9$S$_8$/C	empty nanovoids	366/80	0.5	/	/	36
MoS$_2$/Co$_9$S$_8$/C	nanoboxes	546/100	0.5	/	/	37
Li$_4$Ti$_5$O$_{12}$/MoS$_2$	nanosheets	101/200	/	91	/	38
Na$_3$V$_2$(PO$_4$)$_3$/MoS$_2$	monolithic structure	/	/	70	0.67	39
MoS$_2$/PEO	nanosheets	/	/	225	0.05	40

2. MoS$_2$

Although layered MoS$_2$ shows many advantages over other materials in SIBs, it suffers from the low exfoliation efficiency. Therefore, many researchers are committed to developing various synthetic strategies and have achieved good results.[18,41–43] Bang *et al.*[14] used 1-methyl-2-pyrrolidinone to disperse MoS$_2$ for which the structure is shown in Fig. 1(d). As a result, the as-obtained MoS$_2$ shows higher capacities than the untreated MoS$_2$ in Fig. 1(e). Besides, aiming at confirming that the layered materials can be applied to electrodes in SIBs, Mortazavi *et al.*[44] found through various theoretical calculations that the maximum theoretical capacity (146 mAh g^{-1}) and low average electrode potential (0.75–1.25 V) were obtained due to the phase transition of Na intercalation in MoS$_2$.

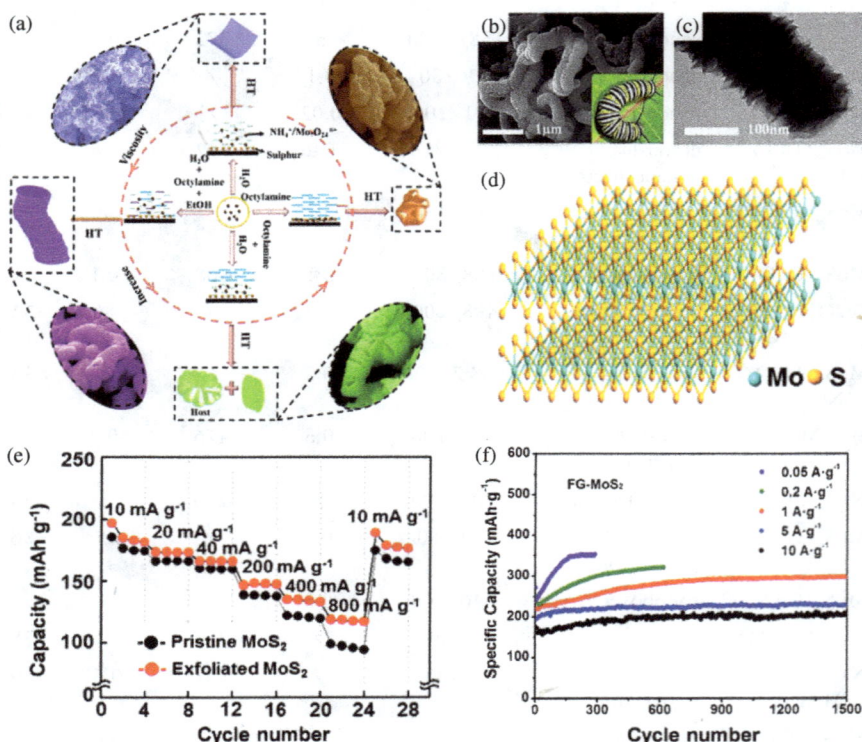

Fig. 1.   (a) The diagram of structure mechanism at various synthetic conditions. (b) The SEM of worm-like MoS$_2$. (c) The TEM of worm-like MoS$_2$. (d) The structure model ofMoS2. (e) The rate capability of different MoS$_2$ samples. (f) Cycle performance of FG-MoS$_2$ at different rates. ((a)–(c)) Reprinted with permission.[15] Copyright 2015, Royal Society of Chemistry. ((d), (e)) Reprinted with permission.[14] Copyright 2014, American Chemical Society. (f) Reprinted with permission.[19] Copyright 2014, Wiley.

Different morphologies of MoS$_2$ have different influences on electrochemical performance of SIBs.[20,45–47] Xu *et al.*[15] synthesized wormlike MoS$_2$ (Figs. 1(b) and 1(c)) by a simple solvothermal method. In Fig. 1(a), the detailed mechanism to prepare the target MoS$_2$ in a different condition can be clearly obtained. As predicted, the target MoS$_2$ as electrode materials in SIBs expressed acceptable capacity (675.3 mAh g^{-1} at 61.7 mA g^{-1}), high cycling stability (83.1% after 80 cycles) and good coulombic efficiency. In general, their research results are very meaningful. They provided a simple and effective method for controlling the synthesis of layered MoS$_2$ and other layered transition metal oxides (TMOs). Besides, Hu *et al.*[19] synthesized flower-like MoS$_2$ which had the function of expanding the layer spacing of the plane by the method of hydrothermal and freeze-drying. As a result, the test of reversible discharge capacities at different rates also testifies this as shown in Fig. 1(f). Analysis of the reasons for such results ultimately attributed this to the fact that the specific structure of the as-prepared sample provided more active sites. In short, materials with high rate performance and long cycling stability are very promising in the field of SIBs.

3. MoS$_2$/Carbon-Based Materials

3.1. *MoS$_2$/carbon materials*

Carbon (C) nanomaterials are one of the most attractive carbon-based materials due to the merits of an excellent conductivity and large surface area. Lu *et al.*[48] used one-step spraying way to prepare MoS$_2$/C microspheres. The obtained composites delivered a good specific capacity (390 mAh g^{-1} with the coulombic efficiency of about 99.8%) as shown in Fig. 2(d). Wang *et al.*[21] fabricated MoS$_2$/C nanospheres which delivered high capacity of 520 mAh g^{-1} at 0.1 C. Similarly, Yang *et al.*[22] used a universal method to get MoS$_2$/C nanospheres which demonstrated long cycles (410 mAh g^{-1} after 1000 cycles at 4 A g^{-1}) in Fig. 2(e). Besides, Park *et al.*[49] prepared MoS$_2$/hierarchical porous carbon (MHPC) composites (Fig. 2(a)) by using a solventless process. The composites delivered good cycle stability and rate capability as shown in Fig. 2(f). Shi *et al.*[23] prepared MoS$_2$/C sandwiched

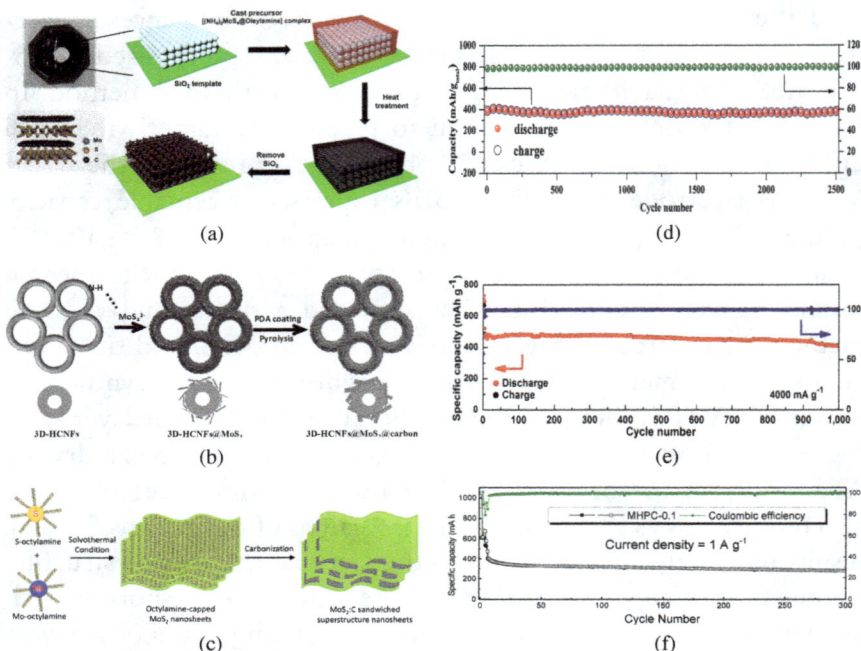

Fig. 2. (a) The preparing process of MHPC composites. (b) The route of the composites multiwalled carbon@MoS$_2$@carbon. (c) The synthesis of MoS$_2$/C nanosheets. (d) The electrochemical properties of MoS$_2$/C: the cycling performance, the coulombic efficiency. (e) The cycling stability of the MoS$_2$/C (f) Cycling performance of the prepared MHPC. ((a), (f)) Reprinted with permission.[49] Copyright 2016, American Chemical Society. (b) Reprinted with permission.[60] Copyright 2016, Wiley. (c) Reprinted with permission.[23] Copyright 2016, Elsevier. (d) Reprinted with permission.[48] Copyright 2016, Wiley. (e) Reprinted with permission.[22] Copyright 2018, Wiley.

nanosheets as shown in Fig. 2(c) which exhibited prominent rate and cycling properties. In addition, there are other research efforts that have made breakthroughs in such composites.[50–53]

To study these materials in depth, many researchers have made efforts.[54–57] Pang *et al.*[58] synthesized N-doped carbon ribbons/MoS$_2$ nanosheets which provided a specific capacitance of 366 mAh g^{-1} at 1 A g^{-1}, coulombic efficiency of initial 75.6% and good cycling stability. Ren *et al.*[59] fabricated carbon cloth@N-doped carbon@MoS2 nanosheets (CC@CN@MoS$_2$), which showed an outstanding rate capability (432.7 F g^{-1}) and excellent cycling stability (265 mAh g^{-1} at

1 A g⁻¹ after 1000 cycles). To improve the performance of composites, Wang et al.[60] synthesized multiwalled C@MoS$_2$@C (the synthesis path as shown in Fig. 2(b)) which achieved an outstanding rate behavior of 817 mAh g⁻¹ at 7000 mA g⁻¹ and a wonderful cycling performance of 747 mAh g⁻¹ after 200 cycles.

3.2. *MoS₂/graphene materials*

It is well-known that graphene materials are widely used in the development of composite materials for their outstanding conductivity.[61–67] To study the ion diffusion problem as the core, Zhao et al.[68] used a facile way to design MoS$_2$-C@ graphene (G) sandwich-type nanostructure which exhibited good capacities, an outstanding reversible capacity (684 mAh g⁻¹) and a remarkable cycling performance as shown in Fig. 3(e). Besides, Geng et al.[25] prepared 1T MoS$_2$@G@MoS$_2$ through hydrothermal for which the process is illustrated in Fig. 3(c). As a result, the as-obtained composites showed a fine reversible capacity (313 mAh g⁻¹, 0.05 A g⁻¹, 200 cycles) and a good rate capability (175 mAh g⁻¹, 2 A g⁻¹). Liu et al.[26] prepared MoS$_2$/G which exhibited a prominent capacity (158 mAh g⁻¹, 1500 cycles, 5 A g⁻¹) as shown in Fig. 3(d). Choi et al.[24] used a one-pot spray pyrolysis method to prepare MoS$_2$/G (Fig. 3(b)) which delivered excellent electrochemical performance. The highest capacity of 480 mAh g⁻¹, especially 88% capacity retentions are obtained in Fig. 3(f). David et al.[69] prepared MoS$_2$/reduced graphene oxide (rGO) by the method of thermal reduction. The as-prepared composites showed charge capacity of 230 mAh g⁻¹ and coulombic efficiency up to about 99%. It is noteworthy that Xie et al.[70] focused on researching the synergistic effect between MoS$_2$ and G. To increase the performance of SIBs, researchers have tried to introduce other substances into the system of MoS$_2$@G.[71] Zhang et al.[72] prepared G@MoS$_2$@SnS$_2$ (Fig. 3(a)) which delivered a discharge capacity of 100 mAh g⁻¹ at 80 mA g⁻¹. Li et al.[73] synthesized sulfur-doped G/MoS$_2$ which provided an outstanding specific capacity up to 587 mAh g⁻¹, a long cycling stability of about 85% retention of original value over 1000 cycles. To determine how to use rGO effectively, many results are worth learning from in the future.[74–77]

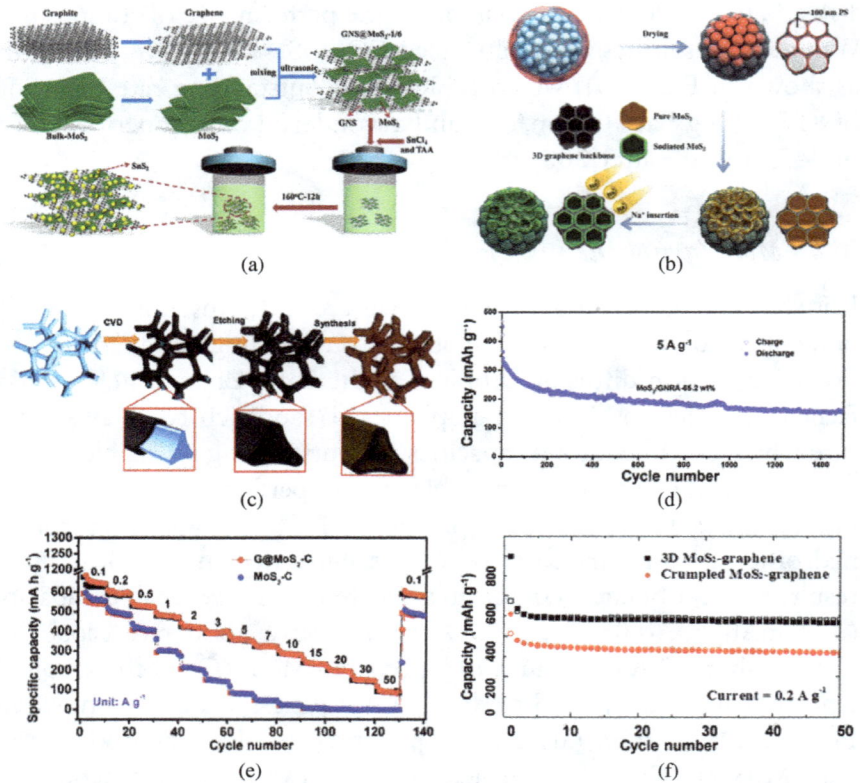

Fig. 3. (a) The synthetic route of the composites. (b) The mechanism of the processing MoS$_2$/G composite (c) Hydrothermal synthesis of the composites. (d) The composites of cycling performance. (e) The rate performance of two samples at different densities. (f) The cycling performances of two samples at 0.2 A g^{-1} (a) Reprinted with permission.[72] Copyright 2017, Elsevier. ((b), (f)) Reprinted with permission.[24] Copyright 2015, Wiley. (c) Reprinted with permission.[25] Copyright 2017, Wiley. (d) Reprinted with permission.[26] Copyright 2016, Elsevier. (e) Reprinted with permission.[68] Copyright 2018, Wiley.

3.3. *Others*

In addition to the above-mentioned carbon-based materials, carbon nanotubes (CNTs) and carbon nanofibers (CNFs) have also been the most popular research hotspots in recent years. Zhang *et al.*[78] prepared MoS$_2$@CNTwhich delivered a specific capacity (480 mAh g^{-1},

0.5 C, 200 cycles) by a hydrothermal strategy. Xu *et al.*[79] synthesized MoS$_2$@CNT which showed a reversible capacity (461 mAh g^{-1} at 500 mA g^{-1} over 150 cycles) and almost 100% coulombic efficiency. Chen *et al.*[80] used a method of a chemical vapour deposition to synthesize MoS$_2$@CNF which showed a reversible capacity (380 mAh g^{-1}) and a good cycling stability after 50 cycles. The results of these studies suggested that these composites had the potential to be expected in SIBs.[81–86]

4. MoS$_2$/Metal Oxides

Various metal oxides as electrode materials for SIBs have achieved increasingly more attention owing to their high specific capacity.[32,87] Ahmed *et al.*[27] prepared HfO$_2$@MoS$_2$ (Fig. 4(b)) which delivered a specific capacity of 636 mAh g^{-1} at 100 mA g^{-1}, compared with the bare MoS$_2$, the composites had a greater electrochemical improvement. TiO$_2$ has the merits of low cost, "green" and low volume expansion in the field of SIBs.[88] Liao *et al.*[28] prepared TiO$_2$/MoS$_2$ which delivered higher rate performance than the pure TiO$_2$ as shown in Fig. 4(e). Only compositing with metal oxides does not greatly improve the properties of MoS$_2$ as electrode material. Kong *et al.*[31] used an applicative way to prepare Fe$_3$O$_4$@MoS$_2$@G as electrode for SIBs (Fig. 4(a)). As a result, the as-obtained composites show a higher discharge capacity (388 mAh g^{-1}, 300 cycles, 100 mA g^{-1}) than that of bare MoS$_2$@G (Fig. 4(d)). Besides, Chen *et al.*[30] prepared MoS$_2$@SnO$_2$@C (the process is shown in Fig. 4(c)) which showed the highest specific capacity among the three materials in Fig. 4(f).

5. MoS$_2$/Metal Sulfides

Metal sulfides are more and more popular for the SIBs owing to the advantages of the outstanding reversible capacity and excellent cycling ability. Choi *et al.*[34] used the way of spray pyrolysis to prepare SnS@MoS$_2$ (Figs. 5(d) and 5(e)) which delivered a discharge capacity of 396 mAh g^{-1} after 100 cycles. Wang *et al.*[35] prepared Ni$_3$S$_2$@MoS$_2$ which exhibited a high reversible specific capacity (568 mAh g^{-1},

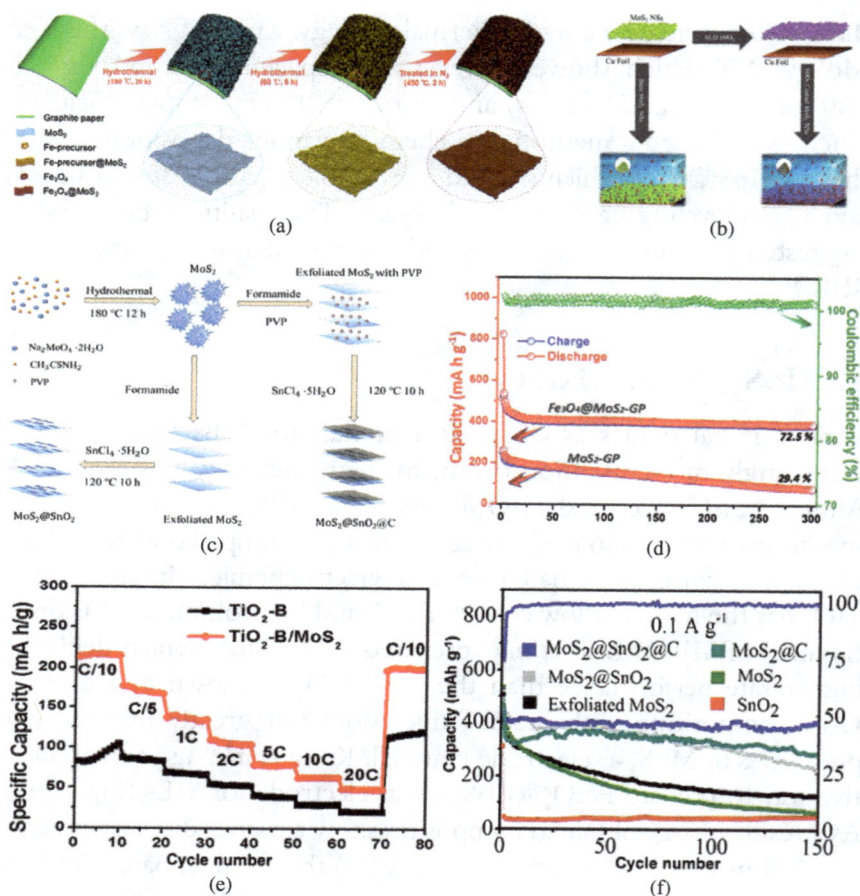

Fig. 4.   (a) The processing for the Fe_3O_4@MoS_2 composite on graphite paper. (b) The process for bare MoS_2 and the HfO_2@MoS_2. (c) The synthesizing of the MoS_2@ SnO_2@C composite. (d) The cycling performance and the corresponding coulombic efficiencies of the two samples. (e) The rate capability of TiO_2 and TiO_2/MoS_2 electrodes. (f) The cycling performance of three samples. ((a), (d)) Reprinted with permission.[31] Copyright 2017, The Royal Society of Chemistry. (b) Reprinted with permission.[27] Copyright 2015, Wiley. ((c), (f)) Reprinted with permission.[30] Copyright 2018, Wiley. (e) Reprinted with permission.[28] Copyright 2016, The Royal Society of Chemistry.

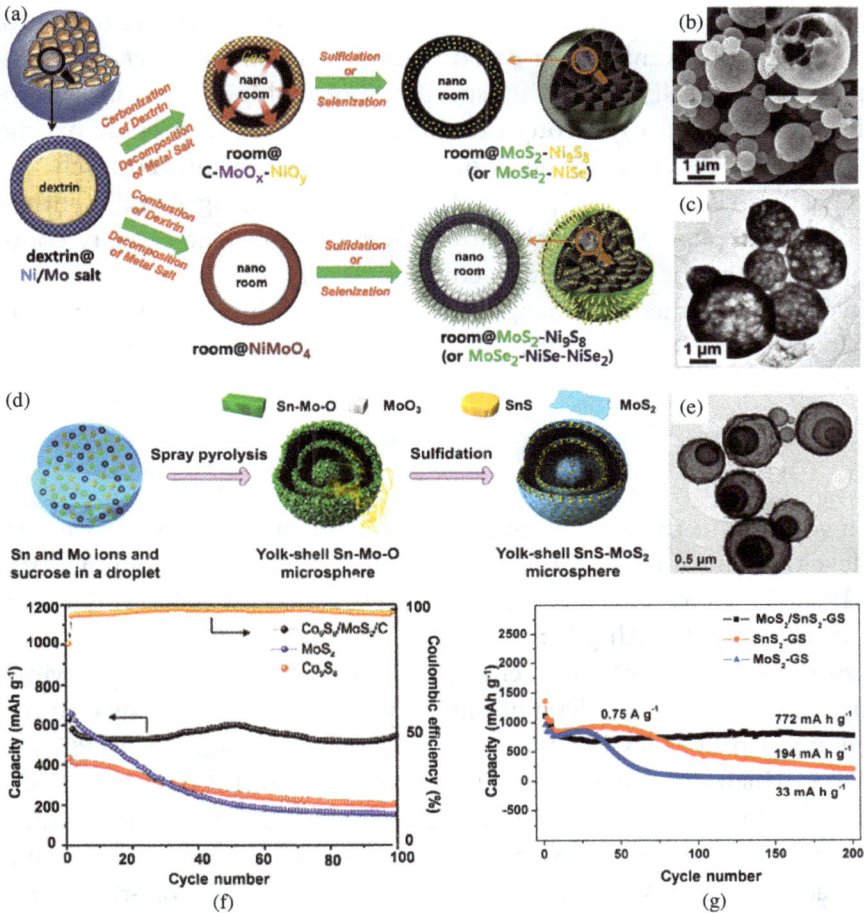

Fig. 5. (a) The formation mechanism of various products. (b) The SEM of MoS_2@ Ni_9S_8@C microspheres. (c) The TEM of MoS_2@Ni_9S_8@C microspheres. (d) The formation mechanism of SnS@MoS_2 composite. (e) The TEM of SnS@MoS2 composite. (g) The cycling performance of three samples. (f) The cycling performance of as-prepared composites. ((a)–(c)) Reprinted with permission.[36] Copyright 2017, The Royal Society of Chemistry. ((d), (e)) Reprinted with permission.[34] Copyright 2015, American Chemical Society. (g) Reprinted with permission.[89] Copyright 2017, American Chemical Society. (f) Reprinted with permission.[37] Copyright 2017, The Royal Society of Chemistry.

200 mA g^{-1}) via an one-step reaction of hydrothermal. Taking into account the advantages of carbon-based materials, Park et al.[36] prepared MoS$_2$@Ni$_9$S$_8$@C (the specific morphologies are shown in Figs. 5(b) and 5(c)) through the method of pilot-scale spray drying (Fig. 5(a)). As a result, the as-obtained composite showed discharge capacities (366 mAh g^{-1}, 80 cycles, 0.5 A g^{-1}). Jiang et al.[89] synthesized MoS$_2$@SnS$_2$@G which shows the most excellent capacity compared with the other two composites (Fig. 5(g)) by one-pot hydrothermal method. Xiang et al.[37] used solvothermal way to obtain MoS$_2$@Co$_9$S$_8$@C which delivered a high capacity (546 mAh g^{-1},100 cycles) in Fig. 5(f).

6. Others

In addition to the various candidate materials summarized above, the development of other materials (Na, BN etc.) have been ongoing for SIBs.[40,90] Xu et al.[38] obtained Li$_4$Ti$_5$O$_{12}$/MoS$_2$ which exhibited a capacity of (91 mAh g^{-1}, 5 C) and a high capacity retention via self-assembly. Wang et al.[91] embedded several materials (polyvinyl pyrrolidone, ethylene diaminetrimolybdate) in MoS$_2$ to study electrochemical properties in SIBs. Zhang et al.[39] obtained Na$_3$V$_2$(PO$_4$)$_3$/MoS$_2$ which showed a high cycling stability (84%, 1000 cycles, 0.5 C) and a reversible capacity (70 mAh g^{-1} at 10 C). Li et al.[40] synthesized poly (ethylene oxide)@MoS$_2$ composites which showed a specific capacity (225 mAh g^{-1} at 50 mA g^{-1}) by the exfoliation–restacking method. The results of these studies shed light on the development of MoS$_2$ composites for SIBs in future.

7. Conclusions and Outlook

In summary, MoS$_2$ and its composites are promising as electrode materials for SIBs. However, the depth of research on these composites for SIBs still faces some challenges. We list the following suggestions for the purpose of trying to summarize the issues that need to be considered in future research.

- Finding the best synthesis strategy. The overall aim is to be green, cost effective and efficient.
- Selecting the appropriate material for compositing with MoS_2. Most of the properties of composites will be better than that of single materials. Therefore, it is particularly important to select materials according to the research needs.
- It is important to regulate the size of the particles of the composite material. According to the study, it was found that most of the better-performing composites have particle sizes controlled at the nanoscale and that the morphology is uniform and regular in shape.

Finally, we hope that this chapter can bridge the research work of the previous years and future research ideas, and provide the basis for the development of MoS_2-based materials for SIBs.

Acknowledgments

This work was supported by the National Natural Science Foundation of China (NSFC-21671170, 21673203 and 21201010), the Top-notch Academic Programs Project of Jiangsu Higher Education Institutions (TAPP), Program for New Century Excellent Talents of the University in China (NCET-13-0645) and the Science and Technology Innovation Foster Foundation of Yangzhou University (2016CXJ010), the Six Talent Plan (2015-XCL-030), Qinglan Project of Jiangsu and Natural science project of Guangling College of Yangzhou University (ZKZD17004). We also acknowledge the Priority Academic Program Development of Jiangsu Higher Education Institutions and the technical support we received at the Testing Center of Yangzhou University.

References

1. B. Li, M. Zheng, H. Xue and H. Pang, *Inorg. Chem. Front.* **3**, 175 (2016).
2. X. Li *et al.*, *Adv. Funct. Mater.* **28**, 1800886 (2018).

3. Y. Wang *et al.*, *J. Mater. Chem. A* **3**, 15292 (2015).
4. X. Jiao *et al.*, *Chem. Sus. Chem.* **11**, 907 (2018).
5. L. Wang *et al.*, *Chem. Eng. J.* **334**, 1 (2018).
6. Y. An, H. Fei, J. Feng, L. Ci and S. Xiong, *Funct. Mater. Lett.* **9**, 1642008 (2016).
7. X. Ma *et al.*, *Funct. Mater. Lett.* **11**, 1850021 (2018).
8. L. Guo *et al.*, *Mater. Technol.* **32**, 592 (2017).
9. B. Chen *et al.*, *Nanoscale* **10**, 34 (2018).
10. T. Wang, S. Chen, H. Pang, H. Xue and Y. Yu, *Adv. Sci.* **4**, 1600289 (2017).
11. F. Xie, L. Zhang, D. Su, M. Jaroniec and S.-Z. Qiao, *Adv. Mater.* **29**, 1700989 (2017).
12. X. Xiong *et al.*, *Energy Environ. Sci.* **10**, 1757 (2017).
13. Z. Hu, Q. Liu, S.-L. Chou and S.-X. Dou, *Adv. Mater.* **29**, 1700606 (2017).
14. G. S. Bang *et al.*, *ACS Appl. Mater. Interfaces* **6**, 7084 (2014).
15. M. Xu *et al.*, *J. Mater. Chem. A* **3**, 9932 (2015).
16. X. Man, L. Yu, J. Sun and S. Li, *Funct. Mater. Lett.* **9**, 1650065 (2016).
17. K. H. Hu, X. G. Hu and C. C. Hu, *Mater. Technol.* **28**, 169 (2013).
18. X. Wang, Y. Li, Z. Guan, Z. Wang and L. Chen, *Chem. — A Eur. J.* **21**, 6465 (2015).
19. Z. Hu *et al.*, Angew. *Chemie Int. Ed.* **53**, 12794 (2014).
20. P. R. Kumar, Y. H. Jung and D. K. Kim, *RSC Adv.* **5**, 79845 (2015).
21. J. Wang *et al.*, *Small* **11**, 473 (2015).
22. Y. Yang *et al.*, *Adv. Mater.* **30**, 1706085 (2018).
23. Z.-T. Shi *et al.*, *Nano Energy* **22**, 27 (2016).
24. S. H. Choi, Y. N. Ko, J.-K. Lee and Y. C. Kang, *Adv. Funct. Mater.* **25**, 1780 (2015).
25. X. Geng *et al.*, *Adv. Funct. Mater.* **27**, 1702998 (2017).
26. Y. Liu *et al.*, *Carbon.* **109**, 461 (2016).
27. B. Ahmed, D. H. Anjum, M. N. Hedhili and H. N. Alshareef, *Small* **11**, 4341 (2015).
28. J.-Y. Liao, B. Luna De and A. Manthiram, *J. Mater. Chem. A* **4**, 801 (2016).
29. W. Ren, W. Zhou, H. Zhang and C. Cheng, *ACS Appl. Mater. Interfaces* **9**, 487 (2017).
30. Z. Chen, D. Yin and M. Zhang, *Small* **14**, 1703818 (2018).
31. D. Kong *et al.*, *J. Mater. Chem. A* **5**, 9122 (2017).
32. D. Xu *et al.*, *Electrochim. Acta* **265**, 419 (2018).

33. K. Zhu *et al.*, *ACS Sustain. Chem. Eng.* **5**, 8025 (2017).
34. S. H. Choi and Y. C. Kang, *ACS Appl. Mater. Interfaces* **7**, 24694 (2015).
35. J. Wang *et al.*, *Nano Energy* **20**, 1 (2016).
36. J.-S. Park and Y. Chan Kang, *J. Mater. Chem. A* **5**, 8616 (2017).
37. J. Xiang and T. Song, *Chem. Commun.* **53**, 10820 (2017).
38. G. Xu *et al.*, *Adv. Funct. Mater.* **26**, 3349 (2016).
39. J. Zhang *et al.*, *Small* **13**, 1601530 (2017).
40. Y. Li *et al.*, *Nano Energy* **15**, 453 (2015).
41. J. Hao *et al.*, *Sci. Rep.* **8**, 2079 (2018).
42. S. H. Woo *et al.*, *Isr. J. Chem.* **55**, 599 (2015).
43. E. Yang, H. Ji and Y. Jung, *J. Phys. Chem. C* **119**, 26374 (2015).
44. M. Mortazavi, C. Wang, J. Deng, V. B. Shenoy and N. V. Medhekar, *J. Power Sources* **268**, 279 (2014).
45. H. He *et al.*, *Nanoscale Res. Lett.* **11**, 330 (2016).
46. J. Su, Y. Pei, Z. Yang and X. Wang, *RSC Adv.* **4**, 43183 (2014).
47. J. Wang, Q. Zhao and J. Chen, *Chinese J. Chem.* **35**, 896 (2017).
48. Y. Lu *et al.*, *Adv. Funct. Mater.* **26**, 911 (2016).
49. S.-K. Park *et al.*, ACS Appl. Mater. *Interfaces* **8**, 19456 (2016).
50. M. Li *et al.*, *RSC Adv.* **7**, 285 (2017).
51. X. Ma *et al.*, J. Mater. *Sci. Mater. Electron.* **29**, 3492 (2018).
52. H. Zhu, F. Zhang, J. Li and Y. Tang, *Small* **14**, 1703951 (2018).
53. Y.-X. Wang *et al.*, *Chem. Commun.* **50**, 10730 (2014).
54. L. Zhao, L. Qi and H. Wang, *RSC Adv.* **5**, 15431 (2015).
55. D. Xie *et al.*, Chem. — *A Eur. J.* **22**, 11617 (2016).
56. Y. Cai *et al.*, *Chem. Eng. J.* **327**, 522 (2017).
57. F. Zheng *et al.*, *Chem. — A Eur. J.* **23**, 5051 (2017).
58. Y. Pang *et al.*, *J. Mater. Chem. A* **5**, 17963 (2017).
59. W. Ren, H. Zhang, C. Guan and C. Cheng, *Adv. Funct. Mater.* **27**, 1702116 (2017).
60. Y. Wang *et al.*, *Small* **12**, 6033 (2016).
61. S. Kalluri *et al.*, *Sci. Rep.* **5**, 11989 (2015).
62. W. Fu *et al.*, *Sci. Rep.* **4**, 4673 (2015).
63. L. Fei *et al.*, *RSC Adv.* **8**, 2477 (2018).
64. J. Xiang *et al.*, *J. Alloys Compd.* **660**, 11 (2016).
65. Y. Teng *et al.*, *Carbon N. Y.* **119**, 91 (2017).
66. X. Jiao *et al.*, *Nanotechnology* **28**, 315403 (2017).
67. Q. Hao, X. Xia, W. Lei, W. Wang and J. Qiu, *Carbon N. Y.* **81**, 552 (2015).

68. C. Zhao *et al.*, *Adv. Mater.* **30**, 1702486 (2018).
69. L. David, R. Bhandavat and G. Singh, *ACS Nano* **8**, 1759 (2014).
70. X. Xie, Z. Ao, D. Su, J. Zhang and G. Wang, *Adv. Funct. Mater.* **25**, 1393 (2015).
71. X. Li, Z. Feng, J. Zai, Z.-F. Ma and X. Qian, *J. Power Sources* **373**, 103 (2018).
72. X. Zhang *et al.*, *Electrochim. Acta* **227**, 203 (2017).
73. G. Li *et al.*, *Adv. Funct. Mater.* **27**, 1702562 (2017).
74. W. Qin, Y. Li, Y. Teng and T. Qin, *J. Colloid Interface Sci.* **512**, 826 (2018).
75. W. Qin *et al.*, *Electrochim. Acta* **153**, 55 (2015).
76. J. Li *et al.*, *Chem. Eng. J.* **332**, 260 (2018).
77. X. Hu *et al.*, *ACS Nano* **12**, 1592 (2018).
78. X. Zhang, X. Li, J. Liang, Y. Zhu and Y. Qian, *Small* **12**, 2484 (2016).
79. X. Xu *et al.*, *J. Mater. Chem. A* **4**, 4375 (2016).
80. C. Chen *et al.*, *Electrochim. Acta* **222**, 1751 (2016).
81. S. Zhang *et al.*, *ACS Appl. Mater. Interfaces* **6**, 21880 (2014).
82. Y. Liu *et al.*, *ACS Nano* **10**, 8821 (2016).
83. C. Zhu, X. Mu, P. A. van Aken, Y. Yu and J. Maier, *Angew. Chemie Int. Ed.* **53**, 2152 (2014).
84. J.-W. Jung *et al.*, *ACS Appl. Mater. Interfaces* **8**, 26758 (2016).
85. X. Xiong *et al.*, *Sci. Rep.* **5**, 9254 (2015).
86. C. Zhao *et al.*, *Nano Energy* **41**, 66 (2017).
87. H. Zhou *et al.*, *Part. Part. Syst. Charact.* **34**, 1700295 (2017).
88. W.-H. Ryu, J.-W. Jung, K. Park, S.-J. Kim and I.-D. Kim, *Nanoscale* **6**, 10975 (2014).
89. Y. Jiang *et al.*, *ACS Appl. Mater. Interfaces* **9**, 27697 (2017).
90. S. Wang, J. Tu, Y. Yuan, R. Ma and S. Jiao, *Phys. Chem. Chem. Phys.* **18**, 3204 (2016).
91. X. Wang, Z. Guan, Y. Li, Z. Wang and L. Chen, *Nanoscale* **7**, 637 (2015).

Chapter 8

Carbon Nanoflakes as a Promising Anode for Sodium-Ion Batteries

Xiaocui Zhu*, S. V. Savilov[†], Jiangfeng Ni*,[‡], and Liang Li*,[§]

The sharp increase in the cost of lithium resource has driven the research on sodium-ion batteries (SIBs) as sodium shares a similar electrochemical property as lithium. Carbonaceous materials are important anodes for rechargeable batteries, but the prevailing graphite only shows a limited activity towards sodium storage. Herein, we demonstrate that carbon nanoflakes serve as an efficient anode material for SIBs, exhibiting a stable capacity of 148 mAh g^{-1} over 600 continuous cycles at 150 mA g^{-1} and an excellent rate capability of 120 mAh g^{-1} at 1500 mA g^{-1}. More importantly, sodium storage in carbon nanoflakes exhibits a pseudocapacitive behavior, possibly due to their larger interlayer spacing and less-ordered structure vs. crystallized carbon.

Keywords: Carbon nanoflake; sodium-ion batteries; anode; pseudocapacitive behavior.

*School of Physical Science and Technology, Center for Energy Conversion, Materials & Physics (CECMP), Jiangsu Key Laboratory of Thin Films, Soochow University, Suzhou 215006, P. R. China.
[†]Chemistry Department, M.V. Lomonosov Moscow State University, Moscow 119991, Russia.
[‡]jeffni@suda.edu.cn
[§]lli@suda.edu.cn

1. Introduction

The over-consumption of fossil fuels causes deficiency of limited resources and environmental pollution. Hence, utilization and deployment of renewable energy becomes a necessity. This further calls on the development of efficient energy storage technologies that can help to stabilize the renewable.[1] Lithium-ion batteries are a typical energy storage technology that can fulfill such a function.[2] However, the surging cost and the limited resource of lithium have driven researchers to consider sodium-ion technology as sodium shares a similar electrochemical nature as lithium.[3,4] At present, sodium-ion batteries (SIBs) are booming, but significant challenges remain before the commercialization of SIBs.[5] One of the problems is to find a suitable and practically applicable anode material.[6–9] So far, researchers have discovered that carbonaceous materials,[10–12] metal elements,[13–15] and metal oxides[16,17] and sulfides[18,19] show interesting sodium-storage properties. Metals such as bismuth[20,21] and antimony[22] can alloy with sodium, and the theoretical capacity is much higher than carbonaceous materials. However, these alloying-type anodes are facing substantial challenges such as high cost and large volume expansion upon sodiation. Metal oxides and sulfides, which store sodium through a conversion mechanism, suffer from poor conductivity and drastic volume changes.[23–25] In general, carbonaceous materials are attractive because they can provide multiple sites for sodium storage. Unfortunately, well-crystallized graphite only exhibits a limited activity toward sodium,[26] while less-ordered expanded graphite show much higher capacity and rate capability.[27]

In this work, we demonstrate that carbon nanoflakes serve as efficient anode material for SIBs. In the potential range of 0.05–2.5 V (vs. Na$^+$/Na), carbon nanoflakes afford a high reversible capacity of 148 mAh g^{-1} at a current density of 150 mA g^{-1} and a superior rate capacity of 120 mAh g^{-1} at 1500 mA g^{-1}. More importantly, carbon nanoflakes exhibit a stable capacity of 148 mAh g^{-1} over 600 continuous cycles at 150 mA g^{-1}, suggesting that they might be a promising anode material for sodium storage.

2. Experiment Details

Carbon nanoflakes were fabricated following our previous procedure.[28] The morphology of carbon nanoflakes was observed using SEM (Hitachi, SU-8010) and TEM (FEI, Tecnai G2F20) equipped with an EMSA/MAS energy dispersive spectroscope. The structure of the samples was characterized by XRD (Rigaku, D/MAX-2000PC) and Raman spectroscopy (Horiba Jobin Yvon, LabRAM HR800). The electrochemical performance of carbon nanoflakes was tested by 2032-type coin cells. Firstly, sodium alginate binder (20 mg) was dispersed in deionized water under stirring for 30 min, followed by adding carbon nanoflakes (170 mg) and acetylene black (10 mg). The mixture of powder was further stirred for 10 h to form a homogeneous slurry. Then, the slurry was cast onto a copper foil using a doctor-blade technique, and vacuum dried at 70°C for 12 h. The loading of active material in the electrode sheet was approximately 1.3 mg cm^{-2}. 1 M $NaPF_6$ dissolved in diglyme is employed as electrolytes, a sodium foil as the anode, and a composite electrode disk ($d = 12$ mm) as the cathode. Cells were galvanostatically charged and discharged on a Land CT2001A battery test system. The cells were first activated by charging and discharging at a rate of 60 mA g^{-1} for five cycles before longterm cycling and rate tests. Cyclic voltammetry (CV) and electrochemical impedance spectroscopy (EIS) were performed on an AUTOLAB PGSTAT302N electrochemical workstation at room temperature (~27°C). CV test was performed with a three-electrode system between 0 and 2.5 V at varied sweep rates. EIS spectra were recorded on cells with the frequency ranging from 100 k Hz to 0.1 Hz and the oscillation voltage was 10 mV. Prior to testing, the cells were left still for at least 10 h.

3. Results and Discussions

Figure 1 shows SEM and TEM images of as-prepared carbon nanoflakes. The carbon material shows a laminar structure and can be easily exfoliated to extremely thin flakes. The XRD pattern shown in Fig. 2(a) exhibits two diffraction peaks at 22.4° and 43.4°,

Fig. 1.   (a, b) SEM and (c, d) TEM images of carbon nanoflakes.

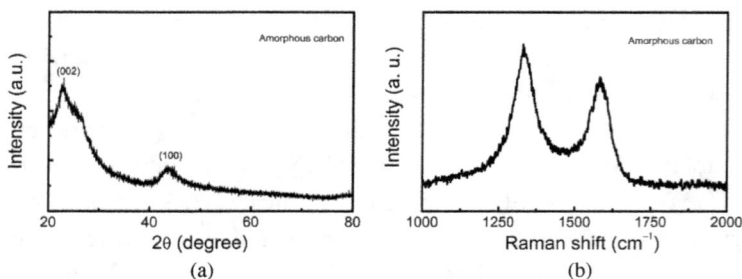

Fig. 2.   Structural characterization of carbon nanoflakes. (a) XRD pattern and (b) Raman spectrum.

corresponding to the (002) and (100) plane of the layered carbon, respectively.[29] The calculated interlayer distance of 0.397 nm for the carbon nanoflake is much larger than that of graphite (0.334 nm), suggesting its potential for sodium storage application. Figure 2(b) shows Raman spectroscopy of carbon nanoflakes, revealing two characteristic bands located at 1329 cm^{-1} and 1581 cm^{-1}. These two bands reflect the well-known defect-induced mode and graphitic mode, respectively.[29,30]

We assess the sodium-storage features of carbon nanoflakes in half coin cells using Na as the counter electrode. Characteristics of sodium storage are indicated by CV measurement. Interestingly, the CV profile manifests a smooth curve without evident redox peaks (Fig. 3(a)). This feature combined with the slope in the galvanostatic curve (Fig. 3(b)) indicates a pseudocapacitive mechanism.[31] A similar phenomenon has also been observed in expanded graphites and some other anode materials.[27] The sample exhibits initial sodiation/desodiation capacities of 209 mAh g⁻¹ and 150 mAh g⁻¹ at 60 mA g⁻¹, respectively. The Coulombic efficiency (CE) in the first cycle is 72%, which gradually increases to 98% in the third cycle, signifying an enhanced reversibility.[31–33] Figure 3(c) shows the cycling stability at a rate of 150 mA g⁻¹. The nanoflake retains a capacity of 148 mAh g⁻¹ over 600 cycles, compared to the initial value of 144 mAh g⁻¹. More importantly, the CE keeps a high value of ~100%. The EIS results

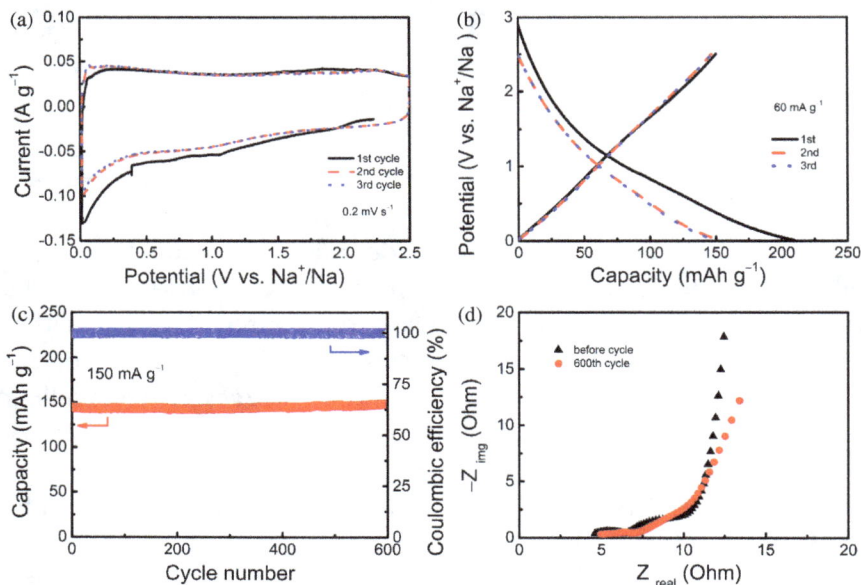

Fig. 3. Electrochemical sodium storage in carbon nanoflakes. (a) CV curves at a sweep rate of 0.2 mVs⁻¹ and (b) Galvanostatic profiles at a rate of 60 mA g⁻¹. (c) Cycling stability at a rate of 150 mA g⁻¹. (d) Impedance spectra before and after 600 cycles.

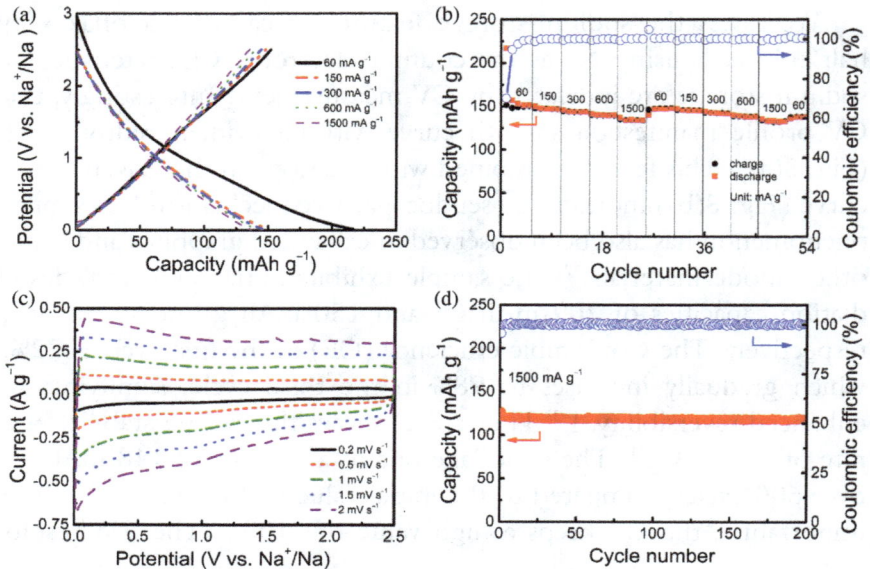

Fig. 4. Rate capability and sodium storage kinetics in carbon nanoflakes. (a) Galvanostatic curves measured at various current rates. (b) Rate cycling capability at various rates. (c) CV curves measured at varied sweep rates. (d) Cycling stability at a high rate of 1500 mA g⁻¹ for 200 cycles.

show little change before and after cycles, indicating a stable reaction kinetic (Fig. 3(d)).[34–36]

As carbon nanoflakes show a pseudocapacitive feature, their capability at high rates was further explored. Figure 4(a) shows the charge–discharge profiles at various current rates. Clearly, neither evident capacity nor voltage decay was observed when the current rate increases from 60 mA g⁻¹ to 1500 mA g⁻¹, suggesting an excellent kinetic feature. In addition, carbon nanoflakes retain a capacity of 134 m Ah g⁻¹ at 1500 mA g⁻¹, which is similar to 148 mAh g⁻¹ at 60 mA g⁻¹ (Fig. 4(b)). The kinetic character is further investigated by CV tests (Fig. 4(c)). A rectangle-like shape is well preserved over various sweep rates, again confirming the pseudocapacitive feature. Hence, the nanoflakes exhibits a stable cycling even at a high rate of 1500 mA g⁻¹ (Fig. 4(d)).

4. Conclusion

In conclusion, we demonstrate that carbon nanoflakes exhibit a superior Na storage performance. The nanoflakes reveal a capacity of 150 mAh g^{-1} at a rate of 60 mA g^{-1} and retain a capacity of 148 mAh g^{-1} over 600 cycles at 150 mA g^{-1}. More importantly, sodium storage in carbon nanoflakes exhibits a pseudocapacitive behavior, as a result of the larger interlayer spacing and less-ordered structure vs. graphitized carbon. It is worth noting that the reversible capacity of carbon nanoflakes is not high. Hence, functionalization and surface treatment may be needed in order to enable a higher degree of sodium storage.

Acknowledgments

The authors are grateful to the financial support of the National Natural Science Foundation of China (Grant Nos. 51672182, 51772197 and 51872192), the Thousand Young Talents Plan, the Jiangsu Natural Science Foundation (Grant Nos. BK20180002 and BK20151219), the Key University Science Research Project of Jiangsu Province (Grant No. 17KJA430013), the 333 High-Level Talents Project in Jiangsu Province, the Priority Academic Program Development of Jiangsu Higher Education Institutions (PAPD), and of the Russian Science Foundation Project (No. 18-13-00217).

References

1. H. Shi *et al.*, *Funct. Mater. Lett.* **10**, 1750076 (2017).
2. K. Xu *et al.*, *Funct. Mater. Lett.* **10**, 1750025 (2017).
3. F. Cheng *et al.*, *Adv. Mater.* **23**, 1695 (2011).
4. N. Yabuuchi *et al.*, *Chem. Rev.* **114**, 11636 (2014).
5. J. Xu *et al.*, *Funct. Mater. Lett.* **06**, 1330001 (2013).
6. S. Fu *et al.*, *Nano Lett.* **16**, 4544 (2016).
7. J. Ni *et al.*, *Adv. Mater.* **29**, 1605607 (2017).
8. J. Ni and L. Li, *Adv. Funct. Mater.* **28**, 1704880 (2018).
9. J. Ni *et al.*, *Adv. Mater.* **28**, 2259 (2016).
10. H. Hou *et al.*, *Adv. Energy Mater.* **7**, 1602898 (2017).

11. J. Yang *et al.*, *Adv. Mater.* **29**, 1604108 (2016).
12. J. Ni and Y. Li, *Adv. Energy Mater.* **6**, 1600278 (2016).
13. D.-H. Nam *et al.*, *ACS Nano* **8**, 11824 (2014).
14. Y. Liu *et al.*, *Adv. Mater.* **11**, 6702 (2015).
15. X. Wang *et al.*, *Adv. Mater.* **26**, 1104 (2015).
16. J. Ni *et al.*, *Adv. Funct. Mater.* **28**, 1707179 (2018).
17. J. Ni *et al.*, *Adv. Mater.* **30**, 1704337 (2018).
18. H. Liang *et al.*, *Nano Energy* **33**, 213 (2017).
19. P. He *et al.*, *Angew. Chem. Int. Ed.* **129**, 12370 (2017).
20. C. Wang *et al.*, *Adv. Mater.* **29**, 35 (2017).
21. C. Wu *et al.*, *Adv. Mater.* **29**, 1604015 (2017).
22. Z. Liu *et al.*, *Energy Environ. Sci.* **9**, 2314 (2016).
23. G. Wang *et al.*, *J. Mater. Chem. A* **1**, 4112 (2013).
24. Y. Zhao *et al.*, *J. Mater. Chem. A* **2**, 13854 (2014).
25. J. Ni *et al.*, *Nano Energy* **34**, 356 (2017).
26. B. Jache and P. Adelhelm, Angew. *Chem. Int. Ed.* **53**, 10169 (2014).
27. Y. Wen *et al.*, *Nat. Commun.* **5**, 4003 (2014).
28. S. V. Savilov *et al.*, *Mater. Res. Bull.* **69**, 7 (2015).
29. P. Lu *et al.*, *Adv. Energy Mater.* **8**, 1702434 (2017).
30. A. C. Ferrari *et al.*, *Phys. Rev. Lett.* **97**, 187401 (2006).
31. J. Ni *et al.*, *Nano Energy* **11**, 129 (2015).
32. J. Liu *et al.*, *J. Mater. Chem. A* **1**, 12879 (2013).
33. Y. Zhao *et al.*, *Nano Res.* **7**, 765 (2014).
34. J. Qiu *et al.*, *J. Membr. Sci.* **297**, 174 (2007).
35. J. Ni *et al.*, *Electrochem. Commun.* **31**, 84 (2013).
36. J. Ni *et al.*, *Electrochim. Acta* **53**, 3075 (2008).

Chapter 9

Phoenix Tree Leaves–Derived Biomass Carbons for Sodium-Ion Batteries

Zengqiang Tian*, Shijiao Sun*,§, Xiangyu Zhao*,†,¶,
Meng Yang*, and Chaohe Xu‡

The biomass carbons with highly developed porosity were prepared by carbonization of phoenix tree leaves (PTLs) followed by activation with KOH. The as-prepared carbons possess large interlayer spacing, abundant oxygen-containing functional groups and high degree of structural disorder. The PTLs derived carbons were firstly employed as anodes for sodium-ion batteries, with the optimized material delivering an initial discharge capacity as high as 602 mAh g^{-1} and a reversible capacity of 134 mAh g^{-1} after 300 cycles at 100 mA g^{-1}. Electrochemical mechanism analysis reveals a capacitive dominating Na^+ storage process.

Keywords: Phoenix tree leaves; biomass carbons; activation; anode; sodium-ion batteries.

---

*College of Materials Science and Engineering, Nanjing Tech University, Nanjing 210009, P. R. China.
†Jiangsu Collaborative Innovation Center for Advanced Inorganic Functional Composites, Nanjing Tech University, Nanjing 210009, P. R. China.
‡College of Aerospace Engineering, Chongqing University, Chongqing 400044, P. R. China.
§sunshijiao@njtech.edu.cn
¶xiangyu.zhao@njtech.edu.cn

1. Introduction

Due to the shortage of fossil fuels and the increasingly serious environment problems, people have realized that it is essential to develop sustainable energies such as wind and solar energy sources. However, due to the instability of the above sustainable energies, large-scale electrochemical energy storage has become a hot spot of the scientific research.[1] Lithium-ion batteries (LIBs) have been widely used in electric vehicles. Nevertheless, they still face big challenges when further employed as large-scale electrochemical energy storage due to the limited lithium mine. Therefore, sodium-ion batteries (SIBs) have attracted people's high attention in recent years because of the natural abundance of Na resources (2.75% of Na compared to 0.065% of Li).[2,3] Moreover, Na has many similar physicochemical and electrochemical characteristics with Li.[4,5] Benefitted from the above characteristics, the SIBs would be expected to replace LIBs for application in large-scale energy storage.

However, the lack of suitable anode materials would limit the commercialization of SIBs.[6] The reason is that the large radius of Na^+ may slow down ion diffusion[7] and produce large volume change. At present, the anode system includes alloys,[8] oxides,[9] sulfides[10] and organic compounds.[11] However, these materials have the common problems of low capacity and poor cycle stability.[12] Hard carbon with higher capacity seems to be the appropriate anode for SIBs.

Hard carbon has many advantages that include low potential for sodium ion insertion (below 0.1 V vs. Na/Na^+), abundant source and considerable capacity (200–300 mAh g^{-1}).[11,13] It is generally supposed that the storage of Na^+ in hard carbon has three ways: (1) adsorption defect sites; (2) Na^+ intercalation into graphitic layers and (3) pore filling.[14,15] However, the major source of hard carbon is pyrolysis of the carbonbased precursor under an inert gas at a very high temperature, which is usually expensive. Therefore, economical ways have to be developed, for instance, the simple carbonization of biomass at a lower temperature.[16]

Natural biomass carbons have attracted great attention because of the hierarchical structure and numerous active sites, such as edge defects and functional groups.[2] Various biomass carbons such as the black fungus,[17] apple biowaste,[18] rape seed shuck,[19] grass,[20] lignin[21] and durian shell[22] have been explored as anodes for SIBs. For example, Wu *et al.*[18] reported that the hard carbon derived from apple biowaste showed a high specific capacity of 245 mAh g^{-1} after 80 cycles at 20 mA g^{-1}. Cao *et al.*[19] demonstrated that the hard carbon made from rape seed shuck exhibited a reversible capacity of 143 mAh g^{-1} after 200 cycles at 100 mA g^{-1}.

Phoenix tree leaves (PTLs) are easily available and have no commercial usage. Although phoenix leaves–derived biomass carbons have been reported to be applied in supercapacitor and water treatment, there are no report on their use for sodium-ion batteries. In this work, the biomass carbons were prepared by carbonization of PTLs followed by activation with KOH. The derived biomass carbons were firstly investigated as anode materials of SIBs. Electrochemical study demonstrated that the biomass carbons carbonized at the optimal temperature of 700°C delivered a reversible capacity of 134 mAh g^{-1} at a current density of 100 mA g^{-1} after 300 cycles.

2. Materials and Methods

Synthesis of biomass carbons: Firstly, the stems were removed from the PTLs. Then the leaves were cleaned by ultrasonic with deionized water and ethanol for 30 min, respectively. Finally, they were dried at 80°C. The clean leaves (10 g) were treated at a designed temperature (700°C, 800°C, 900°C) for 2 h to carbonize in a tubular furnace under constant Ar flow with a heating rate of 5°C min^{-1}. After being cooled down to room temperature, the resultant carbon powder was ground and then mixed with KOH at a mass ratio of 1:2. The obtained mixture was treated at 800°C for 2 h in a tubular furnace under constant Ar flow with a heating rate of 5°C min^{-1}. The mixture was washed with diluted HCl (aq) and deionized water when cooled down to room temperature, then dried under vacuum at 100°C for 12 h. The product was denoted as phoenix leaves–activated carbon-*x* (PLAC-*x*), where *x* refers to the value of the carbonization temperature.

Materials characterization: X-ray diffraction (XRD, Rigaku-SmartLab), fourier transform infrared spectroscopy (FTIR, Thermo Nicolet Nexus 670), Raman spectroscopy (Horiba-Labram HR800) and field emission scanning electron microscopy (FESEM, ZEISS-Ultra55) were used to characterize the structure, morphology and composition of the samples. Nitrogen adsorption–desorption isotherms were measured on a Belsorp–Mini adsorption apparatus. The Brunauer–Emmett–Teller (BET) method was used to calculate the specific surface areas at a relative pressure (P/P0) range from 0.05 to 0.3. The total pore volumes were estimated from the gas adsorbed at a relative pressure P/P0 = 0.99. The micropore volumes were determined by applying Dubinin–Radushkevich (DR) analysis.

Electrochemical measurements: Electrochemical experiments were performed in CR2032-type coin cells. The working electrode was prepared by mixing 80 wt.% active material, 10 wt.% acetylene black, and 10% polyvinylidene fluoride (PVDF) in 1-methyl-2-pyrrolidinone (NMP). The resulting slurry was coated onto a copper foil and dried at 100°C for 12 h under vacuum. The loading of the active materials on the electrode was about 1.76 mg cm^{-2}. Sodium metal was used as the counter electrode and glass fiber was used as the separator. The electrolyte was 1 M NaClO$_4$ dissolved in a mixture of ethylene carbonate (EC) and dimethyl carbonate (DMC) (1:1 by volume) with an addition of 5 wt.% fluoroethylene carbonate (FEC, DoDoChem). Galvanostatic charge and discharge tests were carried out at 100 mA g^{-1} over a voltage range between 0.01 and 3.0 V by using Arbin BT2000 multi-channel battery testing system. Cyclic voltammetry (CV, 0.01–3.0 V) and electrochemical impedance spectroscopy (10^5–0.01 Hz, 5 mV) measurements of the electrodes were performed on a biologic (VMP3) electrochemical workstation.

3. Results and Discussion

The crystalline phase of the samples with different carbonization temperature was identified by XRD, as shown in Fig. 1(a). Two broad diffraction peaks corresponding to the (002) and (101) planes of graphite appeared at around 22.2° and 43.3°, respectively, indicating

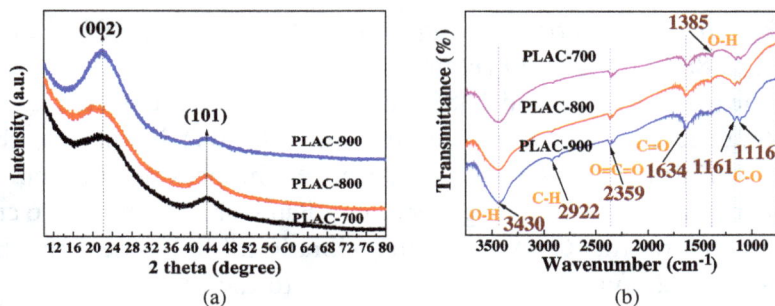

Fig. 1. (a) XRD patterns and (b) FTIR spectra of PLAC-700, PLAC-800 and PLAC-900.

Table 1. Physical parameters for the as-prepared porous carbon materials.

Sample	$^a d_{(002)}$ [nm]	$^b Lc$ [nm]	$^c La$ [nm]	$^d n$	$^e I_D/I_G$	$^f S_{BET}$ [m²g⁻¹]	$^g V_t$ [m³g⁻¹]	$^h V_{micro}$ [cm³g⁻¹]	$^i V_{meso}$ [cm³g⁻¹]
PLAC-700	0.388	1.31	8.66	3.38	2.92	1824	0.89	0.74	0.15
PLAC-800	0.402	1.21	8.97	3.01	2.56	1486	0.71	0.59	0.11
PLAC-900	0.407	1.11	9.77	2.73	3.07	1011	0.60	0.42	0.18

Notes. $^a d_{(002)}$: the interlayer distance of graphitic layers; $^b Lc$: the thickness of graphite-like segments; $^c La$: the average width of graphite-like segments; $^d n$: the average number of graphene layer; $^e I_D/I_G$: the area integral ratio between the D-band and the G-band; $^f S_{BET}$: the specific surface area; $^g V_t$: the total pore volume; $^h V_{micro}$: the micropore volume; $^i V_{meso}$: the mesopore volume.

that the as-prepared carbon materials possess a low degree of graphitization. The $d_{(002)}$ value calculated by Bragg formula based on the (002) peak lies in the range of 0.388–0.407 nm, which represents the interlayer distance of graphitic layers. It is reported that the critical interlayer distance for the Na⁺ insertion is calculated to be 0.37 nm.[23] Hence, the carbons herein are theoretically favorable for Na⁺ insertion. The $d_{(002)}$ value along with the Lc (thickness of graphite-like segments), La (average width of graphite-like segments) and n values (average number of graphene layer) are listed in Table 1. Lc and n decrease, whereas La increases with the increase of the carbonization temperature.

Figure 1(b) shows the FTIR spectra of the as-prepared PLACs. All the samples have almost the same FTIR spectrum. The PLACs exhibit the characteristic absorption peaks at 3430, 2922, 2359, 1634, 1385, 1161 and 1116 cm^{-1}.

The peaks at 3430 and 1385 cm^{-1} are related to the stretching and bending vibrations of the hydroxyl group (−OH) in the adsorbed water during the FTIR test, respectively.[24] The peak at 1634 cm^{-1} is assigned to the C=O stretching vibration.[25] Bands at 1161 and 1116 cm^{-1} correspond to the C–O stretching vibration.[26,27] Besides, the tiny peaks at 2922 and 2359 cm^{-1} are ascribed to the asymmetric stretching vibrations of C–H and CO_2 coordinated with unsaturated surface, respectively.[28,29] Therefore, the FTIR results suggest that the as-prepared PLACs possess a certain amount of oxygen-containing functional groups. It is reported that these oxygen-containing functional groups are favorable for sodium storage.[30]

The PTLs (Fig. 2(a)) were obtained from the campus of Nanjing Tech University. The photograph of the corresponding product PLAC-700 is given in Fig. 2(b). The FESEM observations in Figs. 2(c), 2(e) and 2(g) show that the as-prepared PLACs appear as irregular aggregates with a well-developed pore structure. By careful inspection, each aggregate is composed of nanoparticles. In addition, from the EDX analysis results (Figs. 2(d), 2(f) and 2(h)), we can see that the PLACs contain the C, O, Al, Si and K elements. Among them, the C and O elements account for 97.6–98.8% of the total composition.

Figure 3 displays the Raman spectra of the as-prepared PLACs. All the spectra could be deconvoluted into four Gaussian bands which are referred to as G, D″, D and I bands. The integral intensity ratio of D band to G band (I_D/I_G), representing the disorder degree of carbon, is about 2.56–3.07. The higher values of I_D/I_G demonstrate that the as-prepared biomass carbons possess a high degree of structural disorder, which is consistent with the XRD results above.

N_2 adsorption–desorption measurements were conducted to investigate the textural properties of the PLAC samples. All the adsorption–desorption isotherms (Fig. 3(d)) show a typical Type IV isotherm with H4-type hysteresis, which is the characteristic of the mesoporous material with slit-like pores. Moreover, the steep increase at the lower relative pressure implies that the PLAC samples contain

Fig. 2. Photographs of (a) the Phoenix tree leaf and (b) the corresponding PLAC-700 product. FESEM images and EDX spectra of (c, d) PLAC-700, (e, f) PLAC-800 and (g, h) PLAC-900.

a large fraction of micropores as well. The textural parameters are also summarized in Table 1. The BET specific surface areas of the as-prepared PLACs reach as high as 1011–1824 m² g⁻¹, indicating the well-developed micromesoporosity. By comparing the micropore and

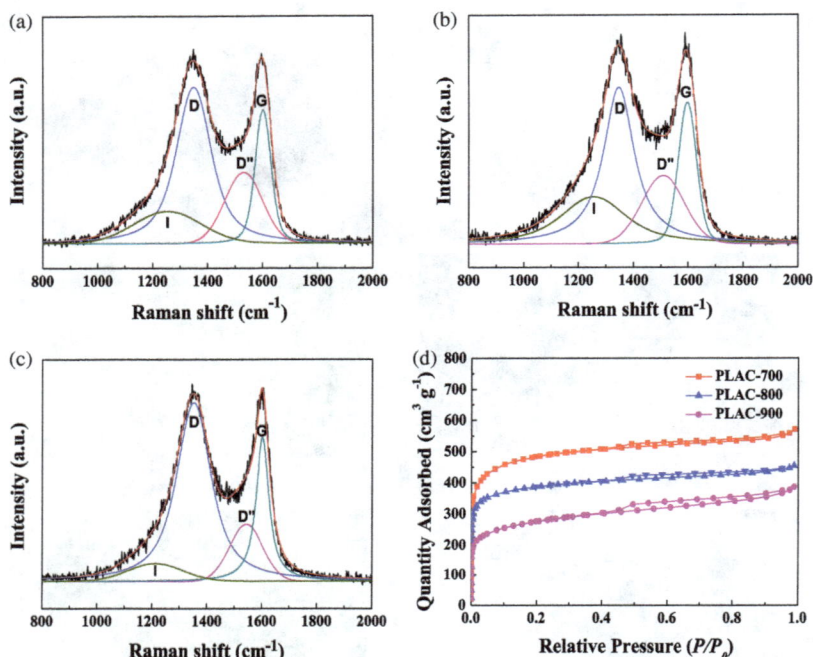

Fig. 3. Raman spectra of (a) PLAC-700, (b) PLAC-800 and (c) PLAC-900 samples and their (d) N_2 adsorption–desorption isotherms.

mesopore volume, we found that micropores are prominent for the carbons since they account for 70–83% of the total pore volumes. Furthermore, the BET specific surface area, total pore volume and micropore volume decrease with the increase of the carbonization temperature.

The electrochemical behaviors of the PTLs derived biomass carbons as anodes of SIBs were tested by the cyclic voltammetry, galvanostatic discharge/charge and electrochemical impedance spectroscopy techniques. Figure S1 (supplementary material) shows the CV curves of the PLAC electrodes at a scan rate of 0.1mV/s. For all the PLAC electrodes, except for the first CV curves, the following CV curves are almost the same and have a regular shape, implying that a capacitive sodium storage behavior is not negligible due to their large surface areas.[31] During the first cathodic process, irreversible reduction peaks are visible. Consistent with the CV results, all the PLAC electrodes show similar voltage profiles, that is a slope in the entire potential region except the first discharge (Figs. 4(a)–4(d)). Different from the well-known two distinct voltage regions of non-porous carbon, the voltage profiles herein are rarely reported and usually encountered for carbon materials with large surface area. The initial discharge capacities of the PLAC electrodes are as high as 602 mAh g^{-1} for the PLAC-700, 496 mAh g^{-1} for the PLAC-800 and 622 mAh g^{-1} for the

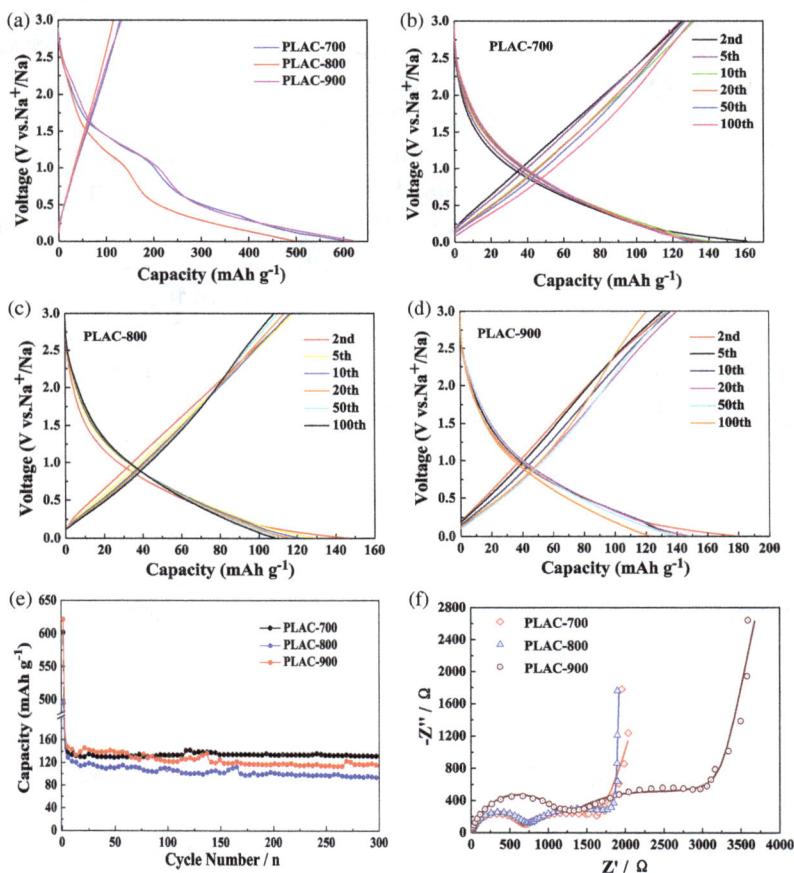

Fig. 4.   (a) Initial galvanostatic charge and discharge curves of PLAC-700, PLAC-800 and PLAC-900 at 0.1 Ag^{-1}; (b), (c) and (d) Subsequent cycles. (e) Cycling performance at current density 0.1 Ag^{-1}; (f) Nyquist plots of the PLAC electrodes measured at the open potential, the dot and the line are related to the experimental and fitting results, respectively.

PLAC-900. Nevertheless, the initial charge capacities of the PLAC electrodes decreased to 115–132 mAh g^{-1} with large irreversible capacity loss. The formation of solid electrolyte interphase (SEI) due to their large surface area may be responsible for the large irreversible capacities of the PLACs.[32] Figure 4(e) shows the discharge capacity as a function of the cycle number for the three samples. After 300 cycles, the discharge capacity maintains 134, 95 and 115 mAh g^{-1} for the PLAC-700, PLAC-800 and PLAC-900, respectively. Compared with the other two electrodes, the best performance of the PLAC-700 electrode may be attributed to its larger specific surface area and total pore volume, which would provide higher electrode–electrolyte contact area and faster ion diffusion. The better reaction kinetics of the PLAC-700 electrode was confirmed by the electrochemical impedance spectroscopy. Nyquist plots (Fig. 4(f)) demonstrate that among the three electrodes, the PLAC-700 electrode possesses the smallest charge-transfer resistance (Table S1, supplementary material).

To reveal the capacitive contribution to the total charge storage, we performed CV tests at different scan rates (Figs. S3(a)–S3(c), supplementary material). The current response at a specific voltage $i(V)$ can be identified as the combined contribution of capacitive effect ($k_1 v$) and diffusion-controlled Na$^+$ insertion process ($k_2 v^{1/2}$)

$$i(V) = k_1 v + k_2 v^{1/2},$$

where k_1 and k_2 are constants, v is the scan rate. Hence, it can be assumed that the current response value has an exponential relationship with the scan rate

$$i(V) = a v^b.$$

Accordingly, $\log i(V) = b \log v + \log a$,
where a and b are constants, with b values ranged from 0.5 (diffusion-controlled contribution) to 1 (capacitive contribution). In our case, the calculated b value for all the three electrodes (Figs. S3(g)–S3(i)) lies between 0.775 and 1, indicating a relatively high capacitive contribution. That is to say, the majority of the capacity originates from surface capacitive effects.

4. Conclusion

In summary, we have successfully prepared the biomass carbons by carbonization of PTLs followed by activation with KOH. Structural characterizations indicate that the PTLs-derived carbons possess large interlayer distances of 0.388–0.407 nm, high degree of structural disorder and well developed micro-mesoporosity. The carbonization temperature has a significant influence on their microstructure. The average width of graphite-like segments increases, while the BET specific surface area, total pore volume and micropore volume decrease with the increase of the carbonization temperature. Accordingly, the biomass carbons carbonized at different temperatures exhibit good Na^+ storage performances. The optimized PLAC-700 delivered an initial discharge capacity as high as 602 mAh g^{-1}, with 134 mAh g^{-1} remained after 300 cycles. Detailed electrochemical mechanism analysis reveals that majority of the capacity originates from surface capacitive effects.

Acknowledgments

This work was supported by the National Natural Science Foundation of China (Grant No. 51602150), the Natural Science Foundation of Jiangsu Province (Grant No. BK20161005), the Priority Academic Program Development of Jiangsu Higher Education Institutions (PAPD) and the Fundamental Research Funds for the Central Universities (No. 201822008).

References

1. J. F. Ni *et al.*, *Funct. Mater. Lett.* **9**, 1650004 (2015).
2. S. Qiu *et al.*, *Adv. Energy Mater.* **7**, 1700403 (2017).
3. J. F. Ni *et al.*, *ACS Energy Lett.* **3**, 1137 (2018).
4. L. F. Xiao *et al.*, *Adv. Energy Mater.* **8**, 1703238 (2018).
5. D. Xie *et al.*, *Adv. Energy Mater.* **7**, 1601804 (2017).
6. S. Liu *et al.*, *Energy Environ. Sci.* **9**, 1229 (2016).
7. J. Xu *et al.*, *Funct. Mater. Lett.* **6**, 1330001 (2013).
8. T. T. Tran *et al.*, *J. Electrochem. Soc.* **158**, A1411 (2011).

9. X. Ma *et al.*, *Funct. Mater. Lett.* **11**, 1850021 (2018).
10. Q. Chen *et al.*, *Adv. Energy Mater.* **8**, 1800054 (2018).
11. N. Zhang *et al.*, *J. Power Sources* **378**, 331 (2018).
12. Y. Y. Zhu *et al.*, *Carbon* **129**, 695 (2018).
13. G. Hasegawa *et al.*, *ChemElectroChem* **2**, 1917 (2015).
14. W. Luo *et al.*, Acc. *Chem. Res.* **49**, 231 (2016).
15. J. Y. Zhan *et al.*, *Nano Energy* **44**, 265 (2018).
16. Y. Y. Zhou *et al.*, *Sci. Bull.* **63**, 146 (2018).
17. X. L. Zhang *et al.*, *Chem. Asian* J. **12**, 116 (2017).
18. L. M. Wu *et al.*, *ChemElectroChem* **3**, 292 (2016).
19. L. Y. Cao *et al.*, *J. Alloy. Compd.* **695**, 632 (2017).
20. F. Zhang *et al.*, *ACS Appl. Mater. Interfaces* **9**, 391 (2017).
21. J. Jin *et al.*, *J. Power Sources* **272**, 800 (2014).
22. G. G. Zhao *et al.*, *J. Mater. Chem.* A **5**, 24353 (2017).
23. Y. L. Cao *et al.*, *Nano Lett.* **12**, 3783 (2012).
24. T. T. Yu *et al.*, *ACS Energy Lett.* **2**, 2341 (2017).
25. L. H. Ai *et al.*, *J. Hazard. Mater.* **198**, 282 (2011).
26. J. Y. Zheng *et al.*, *Appl. Surf. Sci.* **299**, 86 (2014).
27. J. S. Guo *et al.*, *Carbon* **44**, 152 (2006).
28. N. Shukla *et al.*, *J. Magn. Magn. Mater.* **266**, 178 (2003).
29. M. Daturi *et al.*, *Phys. Chem. Chem. Phys.* **1**, 5717 (1999).
30. Y. Y. Shao *et al.*, *Nano Lett.* **13**, 3909 (2013).
31. B. Chao *et al.*, *J. Mater. Chem.* A **4**, 6472 (2016).
32. K. Q. Xu *et al.*, *Funct. Mater. Lett.* **10**, 1650073 (2017).

Chapter 10

Flexible α-Fe$_2$O$_3$ Nanorod Electrode Materials for Sodium-Ion Batteries with Excellent Cycle Performance

Depeng Zhao*, Di Xie*, Hengqi Liu*, Fang Hu*, and Xiang Wu*,†

With the rise of flexible electronics, flexible rechargeable batteries have attracted widespread attention as a promising power source in new generation flexible electronic devices. In this work, α-Fe$_2$O$_3$ nanorods grown on carbon cloth have been synthesized through a facile hydrothermal method as binder-free electrode material. The electrochemical performance measurements show that α-Fe$_2$O$_3$3 nanorods possess high specific capacitance and specific capacity retention of 119% after 100 cycles. The combination of low cost and excellent electrochemical performance makes α-Fe$_2$O$_3$ nanorods promising anode materials for sodium-ion batteries.

Keywords: α-Fe$_2$O$_3$ nanorods; flexible electrode; cycle performance; sodium-ion batteries.

1. Introduction

With the fast growth of the energy shortage crisis and the intense appeal for reducing environmental pollution, alternative energy storage devices have attracted widespread attention.[1-4] Lithium-ion

---
*School of Materials Science and Engineering, Shenyang University of Technology, Shenyang 110870, P. R. China.
†wuxiang05@163.com; wuxiang05@sut.edu.cn

batteries (LIBs) have been widely used in various electronic devices and electric vehicles in various energy storage devices.[5-7] However, large-scale use of LIBs for storing sustainable energy is limited due to high cost and lack of lithium resources. Therefore, sodium-ion batteries are regarded as one of the most promising alternatives, due to the abundance of sodium-containing precursors and the similar chemical properties to that of LIBs.[8,9] Although the electrochemical performance of sodium-ion batteries (SIBs) is slightly less efficient than that of LIBs for the heavier Na^+ and lower operating voltage, the SIBs are still considered as a sustainable and economic choice. Owing to the fact that radius of sodium ions is larger than that of lithium ions, the migration rate of sodium ions in the electrode material is slow, and the intercalation and deintercalation processes are complicated, which cause the low capacity utilization, inferior rate capacity and poor cycling stability for electrode materials during charging and discharging processes.[10,11] Therefore, it remains challenging to exploit the materials with robust structures for accommodating Na^+ and allowing reversible Na^+ intercalation and deintercalation. In order to satisfy the increasing demand for prospective battery technology of SIBs with high cycling performance, large energy and power densities, the research of electrode materials mainly focuses on metal oxides, metal sulfides.

Among various electrode materials, transition metal oxides (TMOs) are considered as promising electrode materials, such as spinel structure of MCo_2O_4 (Ni, Zn and Fe)[12,13] and metal oxide[14] due to the large theoretical specific capacitances originated from multiple valence changes on transition metals. However, α-Fe_2O_3 can be considered as one of the most promising material due to high theoretical capacity, low cost and extensive source and environment friendliness.[15,16] The volume change of α-Fe_2O_3 is large in the charge discharge process, which leads to the exfoliation of the electrodes and the decrease of cycling life. Simultaneously, low electrical conductivity of α-Fe_2O_3 (~10^{-14} S cm^{-1}) greatly limits the practical specific capacitance, especially at high current rates.[17] Slow Na^+ diffusion near α-Fe_2O_3 surface leads to slow electrode kinetics owing to poor electron transfer. To improve the electrochemical performance of α-Fe_2O_3, many efforts have been devoted to preparing various α-Fe$_2$O$_3$ structures and Fe$_2$O$_3$-based hybrid structures.[18,19] Recently, Qiang *et al.* reported the Fe$_2$O$_3$/N-doped ordered mesoporous carbon nanocomposites by multi-step template method, showing an excellent cycle stability of 220 mAh g^{-1} after 300 cycles at a current density of 100 mA g^{-1}.[20] Zhang *et al.* prepared γ-Fe$_2$O$_3$@C nanocomposite which delivers high sodium storage capability and excellent cycle stability of 740 mAh g^{-1} after 200 cycles at 200 mA g^{-1}.[21] Therefore, in order to improve the sodium storage performance of the electrode materials, it is still a great challenge to improve the ion and electron transport in Fe$_2$O$_3$ electrode materials. Herein, we have prepared α-Fe$_2$O$_3$ nanorods through a simple hydrothermal route. The as-prepared products are applied directly as an electrode material for the sodium-ion batteries. It exhibits an initial discharge capacity of 252 mAh g^{-1} at current density of 50 mAh g^{-1} and excellent cycle performance. Thus, α-Fe$_2$O$_3$ nanorods could be excellent candidates for anode material in sodium-ion batteries due to high specific capacity and excellent cycling stability.

2. Materials and Methods

All reagents were of analytical grade and directly used without further purification. Before a typical synthesis, a piece of carbon cloth was cleaned though immersed in 0.3M HNO$_3$ for 10 min, then put into absolute ethanol and deionized water under sonication for 20 min, respectively. After that, the carbon cloth (4 × 4 cm^2) was dried at room temperature. Fe$_2$O$_3$ nanorods were prepared through a typical hydrothermal process. 2 mM ferric nitrate (Fe(NO$_3$)$_2$ · 6H$_2$O), 2 mM sodium sulfate (Na$_2$SO$_4$ · 6H$_2$O), 1.0 g polyvinylpyrrolidone (PVP) were dissolved in 40 ml deionized water under constant magnetic stirring of 40 min for forming a uniform solution. Then, the solution and a piece of cleaned carbon cloth were put into 80 mL Teflon-lined stainless at 100°C for 10 h. After cooled to room temperature, the carbon cloth was taken out and washed with deionized water and absolute ethanol three times, and then the as-prepared products were dried in a vacuum at 60°C overnight. Finally, the as-prepared product was further calcined in 350°C for 2 h at a heating speed of 3°min^{-1} in air. The average mass loading was 1.7 mg cm^{-2}.

The crystallographic structure and phase purity of the asprepared products was measured by power X-ray diffraction analyzer (XRD, 7000, Shimadzu) with Cu Kα radiation (λ = 1.5406 Å). Raman spectra were obtained by a Horiba LabRAM HR Raman spectrometer with excitation wavelength of 532 nm and output power of 5 mW. The specific surface area of the as-prepared products was estimated through Brunauer–Emmett–Teller (BET) analyzer. The pore size distribution was determined with the Barrett–Joyner–Halenda (BJH) method applied to the desorption branch of adsorption–desorption isotherm. The morphology and structure of the as-prepared products were characterized using field emission scanning electron microscope (FE-SEM, Hitachi-4800) and high resolutions transmission electron microscopy (HRTEM) (JEM-2100 PLUS) operated at 200 kV.

The as-prepared electrode materials were cut into the circle plate of 1 cm diameter and were directly acted as the working electrode without adding any binders or conductive agents. The average mass loading of the electrode materials was calculated as 1.7 mg cm^{-2}. CR2032 coin-type cells were assembled in an argon-filled glovebox by using the as-prepared electrode as working electrodes, sodium metal disk was applied to the counter and reference electrode for the sodium-ion batteries. NaPF$_6$ (EC: DMC=1:1) was selected as the electrolyte and glass fiber as the separator. The galvanostatic discharge–charge, rate performance and cycle performance measurement were carried out by LAND CT-2001. The charge/discharge specific capacity of the as-prepared electrode is calculated based on the mass of active materials. Cyclic voltammetry was measured through coin cells on the CHI660E electrochemical workstation (Shanghai, Chenhua Instruments) at a scan rate of 0.1 mVs^{-1} with a voltage range of 0.01–3.0 V. Electrochemical impedance spectra (EIS) were conducted in a frequency range from 0.01 to 100 kHz at 15 mV open circuit potential.

3. Results and Discussion

Figure 1(a) shows XRD pattern of the as-prepared products, it is clearly found that the diffraction peaks can be well indexed to crystal

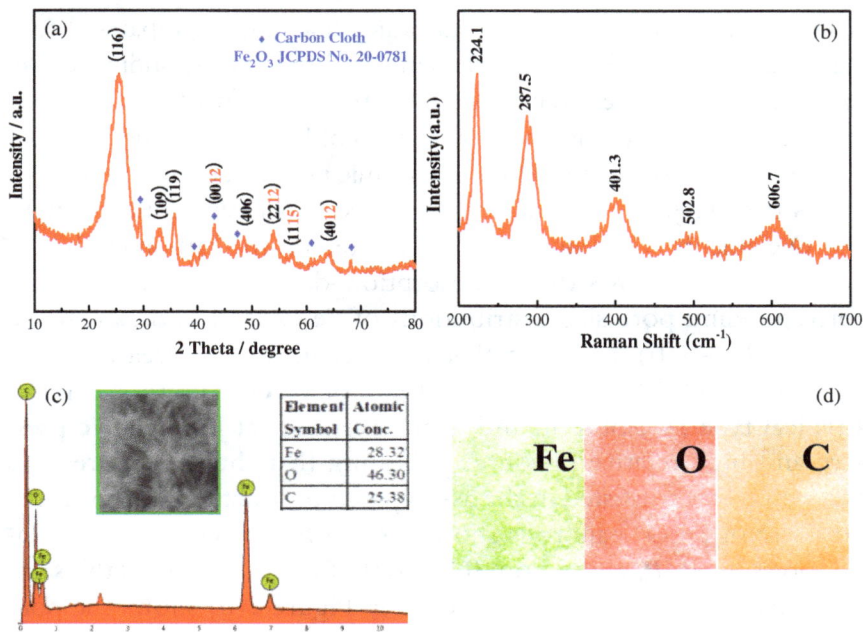

Fig. 1. (a) XRD patterns of the as-prepared products. (b) Raman spectrum of the as-prepared Fe₂O₃ nanorods. (c) EDX spectrum of Fe₂O₃ nanorods, the inset is the corresponding atom ratio. (d) Elemental mappings of Fe, O and C elements.

planes of α-Fe$_2$O$_3$ (JCPDS no. 25-1042). 2θ values of 29.5, 39.5, 43.0, 47.5, 61.9 and 68.4 are well accordance with the diffraction peak of carbon cloth (JCPDS no. 22-1069), which can be attributed to the as-prepared α-Fe$_2$O$_3$ nanorods directly supporting on the carbon cloth. Raman scattering is used to further study structural properties of the as-prepared products, as shown in Fig. 1(b). According to the group theory, Raman-scattering spectroscopy shows five phonon lines. The Raman-scattering peaks are located at 224.1, 287.5, 401.3, 502.8 and 606.7 cm^{-1}, where the peaks at 287.5, 401.3, and 502.8 cm^{-1} can be attributed to the Eg modes.[22] Simultaneously, the peaks centered at 224.1 and 606.7 cm^{-1} are attributed as Raman-active Alg(1) and Alg(2) modes, respectively.[23] It reveals that the as-prepared products are pure α-Fe$_2$O$_3$ phase. The chemical compositions of the α-Fe$_2$O$_3$ nanorods were analyzed through energydispersive X-ray spectrometry

(EDS), as shown in Fig. 1(c), indicating that the as-prepared Fe_2O_3 nanorods consist of Fe and O elements and the corresponding C element can be attributed to carbon cloth. It is well consistent with XRD results. The inset exhibits the atomic ratio of Fe and O is approximatly 2:3, which is close to the theoretical atomic ratio of Fe_2O_3. Figure 1(d) shows the mappings of different elements. It is clearly found that Fe and O elements are all uniformly distributed on the carbon cloth.

Figure 2(a) shows the N_2 adsorption–desorption isotherms and corresponding pore size distributions of Fe_2O_3 nanorods, as shown in Figs. 2(a)–2(b), revealing that the specific surface area of Fe_2O_3 nanorods is 21.48 m^2 g^{-1}. The isotherm of Fe_2O_3 nanorods exhibits a typical type IV isotherm and gently increases at low relative pressure and high relative pressure, indicating that the micropores and the mesopores coexisted in the as-prepared products.[24] However, the isotherm of Fe_2O_3 nanorods shows a small hysteresis loop at a relative pressure P/P_0 range from 0.85 to 0.95, which exhibits the sample is mainly mesoporous in structure.[25] Figure 2(b) shows the pore size distribution curves through the BJH method using the Halsey equation. It can be seen that the average pore size of Fe_2O_3 nanorods is 5 nm and the values of pore distribution can be attributed to the gap between the adjacent small nanostructures. Meanwhile, the pores can effectively improve the permeation of the electrolyte, which benefits for the electrode reaction and electrochemical cyclic performance.

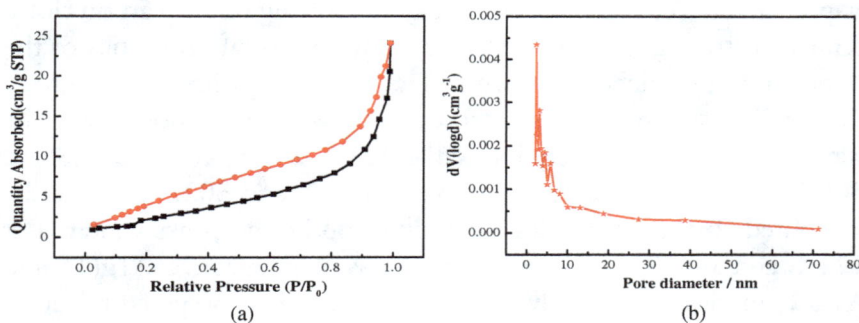

Fig. 2. (a) Nitrogen adsorption–desorption isotherm of Fe_2O_3 nanorods. (b) BJH pore size distribution plot of Fe_2O_3 nanorods.

The morphology and structures of the as-prepared Fe_2O_3 nanorods are characterized by using SEM and TEM. From low magnification SEM images, it can be easily seen that carbon cloth is uniformly covered by many Fe_2O_3 nanorods, as shown in Fig. 3(a). From the high magnification SEM images, as showed in Fig. 3(b), it can clearly obviously that the as-prepared Fe_2O_3 nanorods possess the length of 500 nm and an average diameter of 50 nm. Simultaneously, a large amount of gap are found between the nanorods, which can benefit the transportation of ions between the electrode and the electrolyte. TEM was applied to further characterize the structures of Fe_2O_3 nanorods. Figure 3(c) shows that low magnification TEM images of Fe_2O_3 nanorods, it can be observed a clear spacing between the rods, which is conductive to the transmission of the ions. As can be seen from HRTEM image in Fig. 3(d), the lattice spacing of 0.249 nm, which

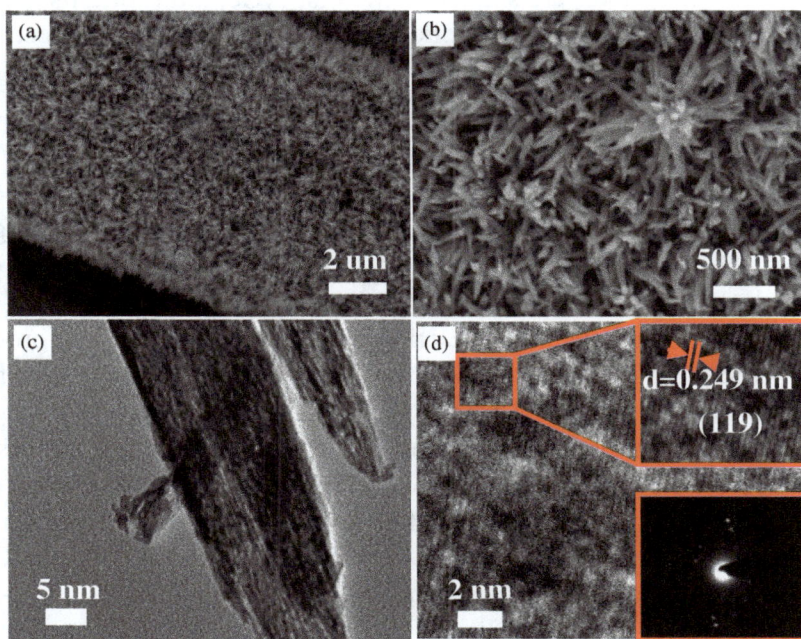

Fig. 3. (a) Low magnification SEM images of Fe_2O_3 nanorods. (b) High magnification SEM images of Fe_2O_3 nanorods. (c) TEM images of the assynthesized Fe_2O_3 nanorods. (d) HRTEM image of Fe_2O_3 nanorods, the inset is the corresponding SAED pattern.

matches well with the (119) plane of Fe_2O_3 spinel structure. The SAED pattern is shown in the inset in Fig. 3(d), indicating that the prepared Fe_2O_3 nanorods are polycrystalline structures. The weak diffraction ring suggests poor crystallinity of as-prepared Fe_2O_3 nanorods, which is well consistent with the XRD analyzation.

To further confirm the electrochemical performance of the as-prepared Fe_2O_3 nanorods, sodium-ion battery was assembled. Figure 4(a) shows CV curves of sodium-ion battery at the voltage of 0.01–3.0 V with scan rate of 0.1 mVs^{-1}. It is clearly seen that an irreversible reduction peak can be observed, which can be attributed to the irreversible reaction of solid electrolyte interface. The pronounced reduction peak occurs at near 0 V, which is due to the reduction of Fe^{3+} to Fe^0. The peaks appear at 0.75 V is considered to be the oxidation from Fe_0 to Fe^{3+} during the first scan.[26,27] In the following scan cycles, the CV curves are similar to the first cycle, indicating its good repeatability due to the network structure. The reversible electrochemical reactions can be described as[28]:

$$Fe_2O_3 + 6Na^+ + 6e^- \leftrightarrow 2Fe + 3Na_2O.$$

Figure 4(b) shows the discharge–charge curves of Fe_2O_3 nanorods at a current density of 50 mA g^{-1}. It is obviously seen that a voltage flat of 1.4 V can be observed, which is ascribed to the formation of solid electrolyte interface and voltage flat at 0.7 V shows that Fe_2O_3 is turn into Fe^{3+} after the initial reaction with Na metal. Fe_2O_3 nanorods shows the specific discharge capacity of 400 mAh g^{-1}. The high specific capacity can be attributed to the Fe_2O_3 nanorods providing a large number of Na^+ insertion active sites during the first discharge process. The irreversible capacity loss can be due to the inevitable formation of solid electrolyte interface (SEI). After several cycles, the voltage plate disappears, indicating that the irreversible reaction between the Fe_2O_3 nanorods and Na metal does not occur after the initial cycle.

The rate capabilities of Fe_2O_3 nanorods were measured at different current densities, as shown in Fig. 4(c). Fe_2O_3 nanorods delivers the discharge capacities of 240, 185, 150, 130, 110, and 210 mAh g^{-1} at current densities of 50, 100, 200, 500, 1000 and 50 mA g^{-1}, respectively. It can be clearly found that after 50 cycles, Fe_2O_3 nanorods still

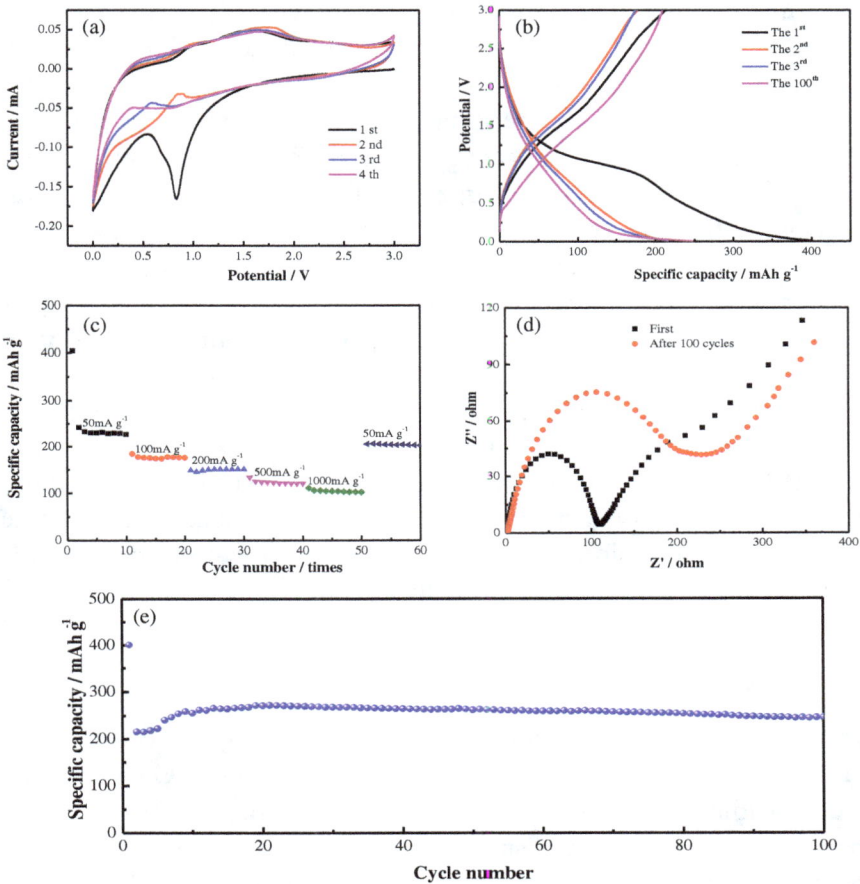

Fig. 4. (a) Cyclic voltammetry curves of Fe$_2$O$_3$ nanorods with a potential window from 0 to 3 V at scan rate of 0.1 m V s^{-1}. (b) Charge–discharge curves of Fe$_2$O$_3$ nanorods for the 1st, 2nd, 3rd and 100th cycles at a specific current of 50 mA g^{-1}. (c) Rate performance of Fe$_2$O$_3$ nanorods. (d) Nyquist plots of Fe$_2$O$_3$ nanorods after 100 times charge–discharge. (e) Cycling performances of Fe$_2$O$_3$ nanorods at a current density of 50 mA g^{-1} for 100 cycles.

showed good stability, indicating that the active materials were protected.

For further study the capacity performance of the electrode materials, electrochemical impedance spectra (EIS) measurement of Fe$_2$O$_3$ nanorods is conducted at a frequency range from 100 kHz to 0.01

Hz, the corresponding Nyquist plots are showed in Fig. 4(d). At high frequency, the intercept at real axis represents a combined internal resistance (Rs), which involve in the intrinsic resistance of active materials, total resistances of the ionic resistance of the electrolyte, and contact resistance at the active materials/current collector interface.[29,30] After 100 cycles, charge transfer resistance increases. This increase in the bulk resistance is similar to a recent detailed EIS study where the influence of electrolyte additives was examined, which suggests that the increase is due to the SEI formation. The cycling performances of Fe_2O_3 nanorods electrodes at a current density of 50 mA g^{-1} are shown in Fig. 4(e). Fe_2O_3 nanorods shows excellent cycle performance, the reversible capacity of 252 mAh g^{-1} is retained after 100 times and the specific capacity retention reached 119% after 100 cycles, the results indicated that excellent reversibility of sodium insertion in Fe_2O_3 nanorods. The increase in capacity after some cycles cycling is mainly due to the reversible formation/decomposition of the oxide and the electrolyte reaction.[31]

4. Conclusion

In summary, we successfully synthesized Fe_2O_3 nanorods via a facile hydrothermal method. The as-prepared Fe_2O_3 nanorods were studied as a dual anode for sodium-ion batteries. Fe_2O_3 nanorods show excellent specific capacity and superior cycle stability performance at current density of 50 mA g^{-1}. The larger reversible capacity may be associated with the synergistic effect between Fe_2O_3 nanorods and short Na$^+$ transfer distance, which leads to enhanced sodiumstorage capacity. The synthesis route provides a potentially promising approach for the development of electrode materials for SIBs.

Acknowledgment

This work was supported by Nature Science Fund of Liaoning province (No. 20170540671).

References

1. Y. X. Zeng *et al.*, *Adv. Energy Mater.* **6**, 1601053 (2016).
2. X. Zhou *et al.*, *Funct. Mater. Lett.* **8**, 1550013 (2015).
3. Y. Zhong *et al.*, *Adv. Energy Mater.* **8**, 1701110 (2018).
4. X. Wu *et al.*, *Nano Energy* **42**, 143 (2017).
5. W. Jiang, *Mater. Res. Bull.* **93**, 303 (2017).
6. Q. Q. Xiong *et al.*, *J. Alloys Compd.* **685**, 15 (2016).
7. L. Wei *et al.*, *Funct. Mater. Lett.* **11**, 1850018 (2018).
8. J. F. Ni *et al.*, *Adv. Funct. Mater.* **28**, 1704880 (2018).
9. S. Li *et al.*, *Electrochimica Acta* **244**, 77 (2017).
10. N. Yabuuchi *et al.*, *Chem. Rev.* **114**, 11636 (2014).
11. S. Wu *et al.*, *Nano Energy* **15**, 379 (2015).
12. D. P. Zhao *et al.*, *New J. Chem.* **42**, 7399 (2018).
13. X. Wu *et al.*, *Nano Energy* **31**, 410 (2017).
14. P. R. Kumar *et al.*, *Electrochim. Acta* **146**, 503 (2014).
15. Y. Jiao *et al.*, *Nano Energy* **10**, 90 (2014).
16. Y. H. Ding *et al.*, *Funct. Mater. Lett.* **6**, 1642008 (2016).
17. Y. X. Zeng *et al.*, *Adv. Energy Mater.* **5**, 1402176 (2015).
18. X. Zheng *et al.*, *Dalton Trans.* **45**, 12862 (2016).
19. X. Liu *et al.*, *Electrochim. Acta* **166**, 12 (2015).
20. Z. Qiang *et al.*, *Carbon* **116**, 286 (2017).
21. N. Zhang *et al.*, *Adv. Energy Mater* **5**, 5 (2015).
22. S. P. S. Porto *et al.*, *J. Chem. Phys.* **47**, 1009 (1967).
23. Y. D. Dong *et al.*, *Vacuum* **150**, 35 (2018).
24. C. Bommier *et al.*, *Carbon* **76**, 165 (2014).
25. R. Ryoo *et al.*, *Adv. Mater.* **13**, 677 (2001).
26. M. Valvo *et al.*, *J. Power Sources* **245**, 967 (2014).
27. Z. L. Jian *et al.*, *Chem. Comm.* **50**, 1215 (2014).
28. B. Philippe *et al.*, *Chem. Mater.* **26**, 5028 (2014).
29. D. P. Zhao *et al.*, *Inorg. Chem. Front.* **5**, 1378 (2018).
30. Y. Jiang *et al.*, *J. Mater. Sci.* **52**, 10950 (2017).
31. Y. Huang *et al.*, *Nano Energy* **41**, 426 (2017).

Chapter 11

Green and Facile Synthesis of Nanosized Polythiophene as an Organic Anode for High-Performance Potassium-Ion Battery

Guifang Zeng*, Yongling An*, Huifang Fei*, Tian Yuan*, Sun Qing*, Lijie Ci*, Shenglin Xiong†, and Jinkui Feng*,‡

In this work, nanosized polythiophene (PTh) was synthesized via a green and facile method. The obtained nanosized PTh was characterized by Fourier transform infrared (FTIR), X-ray diffraction (XRD), Brunauer–Emmett–Teller theory (BET) and scanning electron microscopy (SEM). The results indicated that a nanosized, porous and amorphous PTh was synthesized. As an organic anode material for potassium ion battery, the PTh shows excellent performance with a reversible capacity of 58 mAh g^{-1} at 30 mA g^{-1}.

Keywords: Polythiophene; conducting polymer; anode; potassium ion battery.

*SDU & Rice Joint Center for Carbon Nanomaterials, Key Laboratory for Liquid–Solid Structural Evolution & Processing of Materials (Ministry of Education), School of Materials Science and Engineering, Shandong University, Jinan 250061, P. R. China.
†School of Chemistry and Chemical Engineering, Shandong University, Jinan 250100, P. R. China.
‡jinkui@sdu.edu.cn

1. Introduction

Lithium-ion batteries (LIBs) have captured much attention as a promising energy storage system due to their high energy density and long cycle life.[1-5] However, lithium resource is not only limited but also distributes unevenly in the earth's crust, which results in a high price and restriction for LIBs.[2,3,6,7] Therefore, alternative high-performance energy storage technologies using abundant elements are in urgent demand.[3,8] Sodium and potassium are similar to lithium in physical and electrochemical properties.[9-11] What's more, compared with the barren content of lithium element (0.0017 wt.%), the contents of sodium and potassium elements are as high as 2.36 wt.% and 2.09 wt.% in the earth crust, respectively.[1-3,12-14] Intensive investigations have focused on sodium-ion batteries (SIBs) in the past few years.[3] While potassium-ion batteries (PIBs), which also possess advantages of the low-price and being environmentally friendly, have attracted little attention.[3,14-17] The standard potential of PIBs (−2.93 V for K/K$^+$) is lower than that of NIBs (−2.71V for Na/Na$^+$), which indicates that PIBs have higher theoretical working voltages and energy density.[16] Moreover, K$^+$ ions have a lower charge density due to their large radius which can induce smaller solvated cations, thus PIBs may have higher ionic conductivity and power ability in liquid electrolytes.[2,18] Consequently, PIBs show a promising prospect in the field of battery applications.[16,19]

Currently, to get high-performance PIBs, researchers have put more efforts into the electrode materials.[16] There are mainly three types of materials investigated for the anodes of PIBs: carbon materials, alloy materials and organic materials.[2,20] Carbon materials were introduced for the first time by Wang and his workmates in 2014.[2] They reported that the K$^+$ ions can be reversibly inserted/extracted into/from carbon nanofibers.[21] However, a large volume change was caused due to the large radius of K$^+$ (2.66 Å).[3,21] Alloy anodes such as antimony could deliver a reversible capacity of 650 mAh g^{-1}, which is most approached due to its theoretical capacity (660 mAh g^{-1}).[22,23] Unlike inorganic materials, organic materials have merits of non-toxicity, flexibility and renewability, making them foreground candidates for energy storage.[24]

Conducting polymers, such as Polypyrrole (PPy), Polyaniline (PANI) and Polythiophene (PTh)[25,26] belong to a special family of organic materials. They contain extended π-conjugated systems, in which single and double bonds appear alternately along the polymer chain. They have feasible electrochemical kinetics and high electronic conductivity. Conducting polymers can be used as electrode materials at high-rate via the reversible electrochemical redox (doping/undoping) processes.[27-33] However, as per our knowledge, there are rare reports on conducting polymers as electrode materials for potassium-ion batteries. Herein, we prepared PTh via a facile solvent-free chemical oxidative polymerization method and applied it to potassium-ion batteries. The molecular structure of thiophene monomer and the process of the polymerization reaction are shown in the Fig. 1(a). There are two types of carbon atoms on the thiophene ring: C_α and C_β. Therefore, when the polymerization reaction occurs with the catalyst of $FeCl_3$, there are three types of connections: C_α–C_β connection, C_α–C_β connection and C_β–C_β connection.[27,28,32] PTh connected with C_α–C_α has a higher conductivity than other connections. Some researchers found that it is much easier to form C_α–C_α connection under low temperature conditions. For example, the electronic conductivity of PTh polymerized at 0°C was five times higher than that polymerized at 55°C. So the polymerization temperature plays an important role in the polymerization reaction.[27,28,32] For potassium/PTh half cell, the discharge–charge mechanism is shown in Fig. 1(b). The neutral PTh

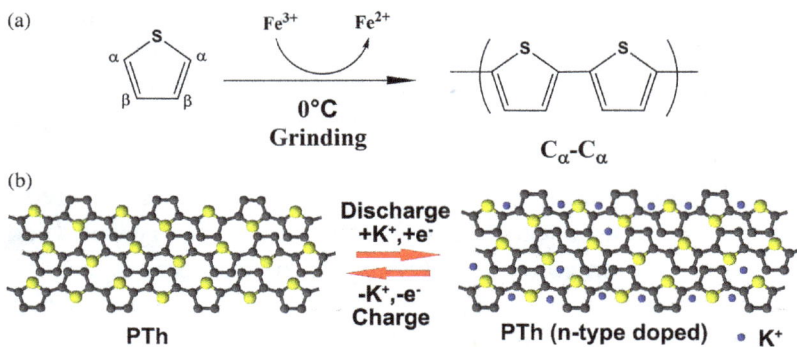

Fig. 1. (a) Synthesis mechanism of PTh, (b) charge–discharge mechanism of PTh/K half cell.

gained both electrons and K^+ ions, which is also named as *n*-type doping. The reaction is reversed when cells were charged.[34,35]

In this work, we probed that PTh can be used as anode for PIBs in the carbonate electrolyte for the first time. It is found that PTh could deliver a reversible capacity of 58 mAh g^{-1} at the rate of 30 mA g^{-1}. The excellent performance benefits from the typical layered, porous and the electronic conductive properties of PTh, which can give channels for both ions and electrons.

2. Materials and Methods

A 0.4 mol $FeCl_3$ was prepared in an agate mortar as catalyst, then 0.1 mol thiophene was slowly added. At the same time, the pestle was used to vigorously grind the mixture for 20 min. After that, the mixture was added to ethanol solution and stirred for 4 h. The mixed solution was first centrifuged and washed twice by ethanol to remove the thiophene monomer. The obtained deposit was added to HCl solution (3 mol/L) to remove the $FeCl_3$ overnight at 0°C. Then it was centrifuged and washed by HCL solution twice. Finally, it was dried in the oven at 80°C to get a fine powder.

Fourier transform infrared (FTIR) spectra of PTh were recorded by a NICOLET AVATAR 360 FT-IR spectrometer from 500 to 1800 cm^{-1} with KBr pellets. X-ray using a Rigaku Dmaxrc diffractometer was employed to analyze the crystal phase of PTh at the scanning speed of 10°/min from 5° to 90°. The surface morphology and structure of PTh were characterized by FESEM (using SU-70 field emission scanning electron microscopy). The Brunauer–Emmett–Teller theory (BET, ASAP 2020) was used to test the porous property of PTh. To test its electrochemical performance, the slurry consists of active material (PTh, 60 wt.%), carbon black (20 wt.%) and carboxy methyl cellulose sodium (CMC) (20 wt.%), which then covered the copper foil current collector. The foil coated with slurry was dried at 70°C under vacuum condition overnight. The diameter of each piece was 14 mm and the mass of active material was about 1–2 mg. The potassium metal was used as the reference electrode and the counter electrode of the PIBs.[3] Glassy-fiber was employed as the separator.

The composition of the electrolyte contained 1 mol KFSI dissolved in a mixture of diethyl carbonate/ethylene carbonate (DEC/EC, v/v = 1:1). All cells were assembled using 2016 coin-type in a glovebox full of argon. A galvanostatic programmable battery charger was used to test the galvanostatic charge/discharge of cells in the voltage range of 0–3 V (versus K/K⁺) measured at room temperature.

3. Results and Discussion

Figure 2(a) shows the FTIR spectrum of PTh sample from 500 to 1800 cm⁻¹. The absorption peaks at 788, 1033, 1132, and 1490 cm⁻¹ confirmed that PTh has been synthetized via the chemical method.[27,28,36,37] The PTh chain is mainly connected by C_α–C_α connection.[28] The typical band at 788 cm⁻¹ is the C_β-H out-of-plane bending vibration of the 2,3,5-disubstituted thiophene ring, which indicates C_α–C_α connection of thiophene monomer has generated with higher conductivity.[27,36,37] The peaks at 694 cm⁻¹ and 1033 cm⁻¹ are the C–S asymmetric and symmetric stretching vibration bands, respectively. The vibration band at 1132 cm⁻¹ is the C–H bending vibration in the thiophene ring, while the band at 1490 cm⁻¹ is the C–H stretching vibration (aromatic stretching band) in the thiophene ring. The C–C stretching vibration appeared at 1213 cm⁻¹ and 1328 cm⁻¹. The absorption vibration band at 1433 cm⁻¹ is assigned to the C=C stretching vibration band.[27,36] The XRD pattern of as-prepared PTh is shown in Fig. 2(b). There is a relatively narrow peak centered at $2\theta = 20°$, which indicates the degree of crystallization is low and the crystal structure of PTh is a typical amorphous structure.[37] Figure 2(c) shows the nitrogen adsorption–desorption isotherms with a BET surface area of 16 m² g⁻¹. It can be seen from Fig. 2(d) that the as-synthesized PTh has a wide pore distribution. The FESEM image of PTh polymer is presented in Figs. 2(e) and 2(f). The PTh powder is mainly presented in an aggregate by fine PTh nanoparticles with a diameter of 10–30 nm, which can give channels for electrons to pass through.[27] Each particle is connected by a coral structure, which may be favorable for electrode reactions due to the effective surface area and ionic channels.

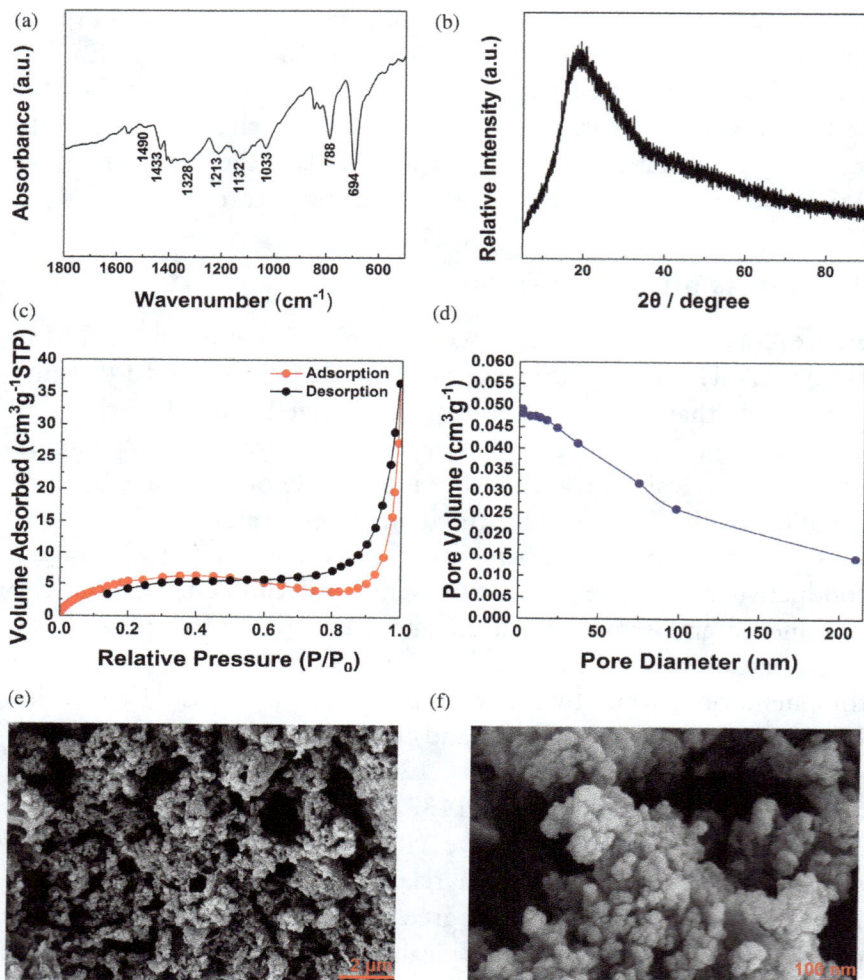

Fig. 2. (a) FTIR spectra of PTh, (b) XRD pattern, (c) the nitrogen adsorption/desorption isotherms, (d) pore size distributions of PTh, (e) FESEM image of low magnification, and (f) FESEM image of high magnification.

The 1st, 2nd, 10th, 50th and 80th charge–discharge curves of PTh anode are shown in Fig. 3(a) between 0 V and 3 V at a rate of 30 mA g^{-1}. In the first cycle, a discharge capacity of 76 mAh g^{-1} and a reversible charge capacity of 58 mAh g^{-1} were obtained, the irreversible capacity is well known owing to the formation of SEI.[3] In the

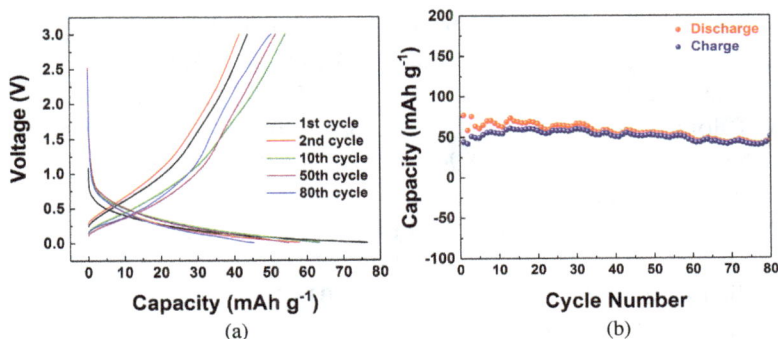

Fig. 3. Electrochemical performance of PTh as the anodes for PIB: (a) The 1st, 2nd, 10th, 50th and 80th charge–discharge voltage profiles at 30 mA g^{-1}. (b) Cycling performance at 30 mA g^{-1}.

second cycle, the coulombic efficiency is greatly improved, which may be due to the formation of stable SEI films. Figure 3(b) shows a discharge capacity of 56 mAh g^{-1} on an average and a discharge capacity of 45 mAh g^{-1} is achieved after 100 cycles. Meanwhile, the curves of charge–discharge are highly overlapped, indicating a good cycling stability.

4. Conclusion

In conclusion, porous PTh nanopowder is synthesized by a green and facile chemical oxidative polymerization method and used as anode materials for PIBs for the first time. The PTh delivers a reversible capacity of 58 mAh g^{-1} and a good capacity retention even after 80 cycles.

References

1. I. Sultana *et al.*, *Chem. Commun.* **52**, 9279 (2016).
2. X. Zou *et al.*, *Chem. Commun.* **19**, 26495 (2017).
3. Y. An *et al.*, *Chem. Commun.* **53**, 8360 (2017).
4. J. Ni and L. Li, *Adv. Funct. Mater.* **28**, 1704880 (2018).
5. B. Li *et al.*, *Adv. Funct. Mater.* **27**, 1605784 (2017).

6. Y. Tang *et al.*, *Adv. Mater.* **26**, 6111 (2014).
7. S. Wang *et al.*, *Adv. Energy Mater.* **6**, 1502217 (2016).
8. S. Zheng *et al.*, *Adv. Energy Mater.* **7**, 1602733 (2017).
9. V. Palomares *et al.*, *Energy Environ. Sci.* **5**, 5884 (2012).
10. J. Ni, L. Li and J. Lu, *ACS Energy Lett.* **3**, 1137 (2018).
11. N. Jabeen *et al.*, *ACS Appl. Mater. Interfaces* **8**, 33732 (2016).
12. N. Yabuuchi *et al.*, *Chem. Rev.* **114**, 11636 (2014).
13. J. Ni *et al.*, *Adv. Mater.* **30**, 1704337 (2018).
14. Z. Jian *et al.*, *Adv. Energy Mater.* **6**, 1501874 (2016).
15. J. Zhao *et al.*, *Adv. Funct. Mater.* **26**, 8103 (2016).
16. B. Ji *et al.*, *Adv. Mater.* **29**, 1700519 (2017).
17. H. Xia *et al.*, *J. Mater. Chem.* A **3**, 1216 (2015).
18. S. Komaba *et al.*, *Electrochem. Commun.* **60**, 172 (2015).
19. W. Luo *et al.*, *Nano Lett.* **15**, 7671 (2015).
20. N. Jabeen *et al.*, *Adv. Mater.* **29**, 1700804 (2017).
21. Y. Liu *et al.*, *Nano Lett.* **14**, 3445 (2014).
22. W. D. McCulloch *et al.*, *ACS Appl. Mater. Interfaces* **7**, 26158 (2015).
23. J. Ni *et al.*, *Nano Energy* **34**, 356 (2017).
24. T. B. Schon, B. T. McAllister, P. F. Li and D. S. Seferos, *Chem. Soc. Rev.* **45**, 6345 (2016).
25. A. F. Diaz *et al.*, *J. Electroanal. Chem. Interfacial Electrochem.* **129**, 115 (1981).
26. J. Desilvestro, W. Scheifele and O. Hass, *J. Electrochem. Soc.* **139**, 2727 (1992).
27. L. Liu *et al.*, *React. Funct. Polym.* **72**, 45 (2012).
28. F. Wu *et al.*, *J. Phys. Chem.* C **115**, 6057 (2011).
29. S. R. Sivakkumar and D.-W. Kim, *J. Electrochem. Soc.* **154**, A134 (2007).
30. F. Tian *et al.*, *Mater. Chem. Phys.* **127**, 151 (2011).
31. M. G. Kanatzidls, *Chem. Eng. News* **68**, 36 (1990).
32. K. S. Ryu *et al.*, *Mater. Chem. Phys.* **84**, 380 (2004).
33. J. K. Feng *et al.*, *J. Power Sources* **177**, 199 (2008).
34. C. Zhang *et al.*, *Adv. Funct. Mater.* **28**, 1705432 (2017).
35. X. Li *et al.*, *Adv. Funct. Mater.* **28**, 1800886 (2018).
36. D. Xu, P. Wang and R. Yang, *Ceram. Int.* **43**, 7600 (2017).
37. N. Ballav and M. Biswas, *Synth. Met.* **142**, 309 (2004).

Chapter 12

Nitrogen-Doped MnO_2 Nanorods as Cathodes for High-Energy Zn-MnO_2 Batteries

Yalan Huang*,[†], Wanyi He[†], Peng Zhang*,[‡], and Xihong Lu[†,§]

The development of manganese dioxide (MnO_2) as the cathode for aqueous Zn-MnO_2 batteries is hindered by poor capacity. Herein, we propose a high-capacity MnO_2 cathode constructed by engineering it with N-doping (N-MnO_2) for a high-performance Zn-MnO_2 battery. Benefiting from N element doping, the conductivity of N-MnO_2 nanorods (NRs) electrode has been improved and the dissolution of the cathode during cycling can be relieved to some extent. The fabricated Zn-N-MnO_2 battery based on the N-MnO_2 cathode and a Zn foil anode presents a real capacity of 0.31 mAh cm^{-2} at 2 mA cm^{-2}, together with a remarkable energy density of 154.3 Wh kg^{-1} and a peak power density of 6914.7 W kg^{-1}, substantially higher than most recently reported energy storage devices. The strategy of N doping can also bring intensive interest for other electrode materials for energy storage systems.

*School of Environment and Civil Engineering, Guangdong Engineering and Technology Research Center for Advanced Nanomaterials, Dongguan University of Technology, Dongguan 523808, P. R. China.
[†]MOE of the Key Laboratory of Bioinorganic and Synthetic Chemistry, The Key Lab of Low-Carbon Chem & Energy Conservation of Guangdong Province, School of Chemistry, Sun Yat-Sen University, Guangzhou 510275, P. R. China.
[‡]zhangpeng@dgut.edu.cn
[§]luxh6@mail.sysu.edu.cn

Keywords: Nitrogen doping; Zn-MnO_2 batteries; MnO_2 nanorods; high-energy.

1. Introduction

The ever-growing demands of next-generation electronics such as rollup displays, implantable medical devices, and electric vehicles call for the urgent development of energystorage devices.[1-6] Advanced battery technologies with high safety and low cost are highly desirable for applications in those consumer electronics,[7,8] among which lithium-ion batteries (LIBs) are most widely used due to their high energy density.[9,10] However, the large-scale application of LIBs is limited by their safety issues and the scarceness of Li resources.[11-15] Rechargeable aqueous batteries have been considered as compelling alternatives to stationary grid-level storage of renewable energies, among which aqueous zinc ion batteries have attracted extensive attention owing to their high safety, low cost, and eco-friendliness.[16-19] In this regard, aqueous Zn-MnO_2 batteries, identified by good safety, cost effectiveness, high output voltage (~1.5 V), and high capacity, are emerging as one of the most promising candidates.[20-22] Recently, considerable efforts have been devoted to rechargeable aqueous Zn-MnO_2 batteries, and much vital progress has been made.[23-25] For instance, Yu *et al.* demonstrated a fiber-shaped Zn-MnO_2 battery reaching a discharge capacity of 0.29 mAh cm^{-1} at a low current density of 0.07 A g^{-1}.[23] Wang and his coworkers assembled an alkaline Zn-MnO_2 battery using MnO_2/CNTs as a cathode, which achieved 236 mAh g^{-1} at 0.3 mA cm^{-2}.[25] Despite many merits of Zn-MnO_2 batteries, they still have some drawbacks. For one thing, alkaline electrolytes are widely used, rendering poor reversibility of Zn-MnO_2 batteries.[26-28] Recently, Liu and his co-workers exploited a mild aqueous $ZnSO_4$ electrolyte with $MnSO_4$ addictive, which is able to reduce the accumulation of the byproducts from both cathode and anode, boosting the cyclic performance of the battery.[29] Additionally, an aqueous pouch of Zn-MnO_2 battery based on $Zn(CF_3SO_3)_2$ electrolyte with $Mn(CF_3SO_3)_2$ additive delivered a total energy density of

75.2 Wh kg^{-1}.[24] However, MnO_2 cathode is hindered by the poor actual capacity and rapid capacity fading, caused by its inferior electroconductivity and structural instability.[30-32] On account of these problems, Zn-MnO_2 batteries always suffer from relatively low capacity and poor cycling durability, hardly satisfying the critical demands of next-generation electronics.

Herein, we developed a high capacity aqueous Zn-MnO_2 battery using the N-doped MnO_2 nanorods (NRs) as the cathode material. The capacity of Zn-MnO_2 battery has been improved by introducing N element into MnO_2 (noted as N-MnO_2) via annealing under NH_3 atmosphere, reaching 0.31 mAh cm^{-2} at a current density of 2 mA cm^{-2} in aqueous electrolyte. Besides, the aqueous Zn-MnO_2 battery achieved an outstanding energy density of 154.3 Wh k g^{-1} at a current density of 2 mA cm^{-2}, accompanied by a peak power density of 6914.7 W kg^{-1} at a current density of 15 mA cm^{-2}. All these results indicate that the method of N doping is an efficient way to improve the capacity of MnO_2.

2. Methods

The schematic illustration of synthetic process of N-MnO_2 is shown in Fig. 1(a). Briefly, N-MnO_2 NRs grown on carbon cloth were synthesized via two-step process. MnO_2 NRs were firstly grown on a carbon cloth substrate through template-free anodic electrodeposition method. At first, carbon cloths with the area of 2.0 × 3.0 cm² were cleaned with ethanol and distilled water by ultrasound, and then dried in drying oven. Later, the anodic electrodeposition was conducted in a two electrode cell using carbon cloth as working electrode and carbon electrode as counter electrode. Anodic electrodeposition was conducted at a constant current density of 0.4 mA cm^{-2} in a solution consisting of 0.02 M manganese acetate ($MnAc_2$) and 0.01 M ammonium acetate (NH_4Ac) at 70°C for 3 h. After that, MnO_2 NRs with the diameter of ~50–100 nm were uniformly coated on the carbon cloth (noted as U-MnO_2, Fig. 1(b)), and the mass loading of MnO_2 NRs was around 2.56 mg cm^{-2}. The N-MnO_2 NRs were obtained by annealing the as-prepared U-MnO_2 NRs in air at 300°C for 1 h

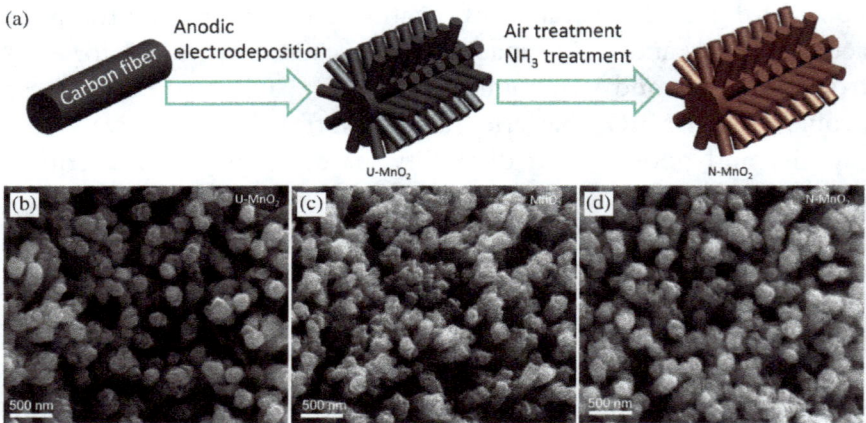

Fig. 1.   (a) Schematic diagram illustrates the growth process for preparing N-MnO$_2$ NRs on carbon cloth substrate. (b–d) SEM images of MnO$_2$-Untreated NRs, MnO$_2$ (heating treated by air) and N-MnO$_2$ on a carbon cloth.

(noted as MnO$_2$) and then in NH$_3$ (flow rate: 100 sccm) at 100°C for 1 h. From the scanning electron microscopy (SEM) images (Figs. 1(c) and 1(d)), the morphologies of MnO$_2$ and N-MnO$_2$ NRs were well preserved during thermal treatments, forecasting the structure of the MnO$_2$ NRs was relatively stable.

3. Results and Discussion

To identify the possible phase transformation, X-ray diffraction (XRD) analysis was performed. As shown in Fig. 2(a), except for the peak of carbon, the diffraction peaks of MnO$_2$ NRs are well in agreement with β-MnO$_2$ (JCPDS# 24-0735). Additionally, the XRD patterns are well consistent between MnO$_2$ and N-MnO$_2$ NRs, suggesting that there is no phase transformation of the MnO$_2$ NRs after thermal treatment at NH$_3$ atmosphere. Figure 2(b) shows the Raman spectra of MnO$_2$ and N-MnO$_2$ NRs, in which the Raman band at 651 cm^{-1} for the MnO$_2$ sample can be attributed to the three major Mn-O stretching vibrations of the [MnO$_6$] group in MnO$_2$, proving that the sample is pure MnO$_2$ after thermal treatment in air.[33] The characteristic Raman peaks of N-MnO$_2$ sample are almost consistent

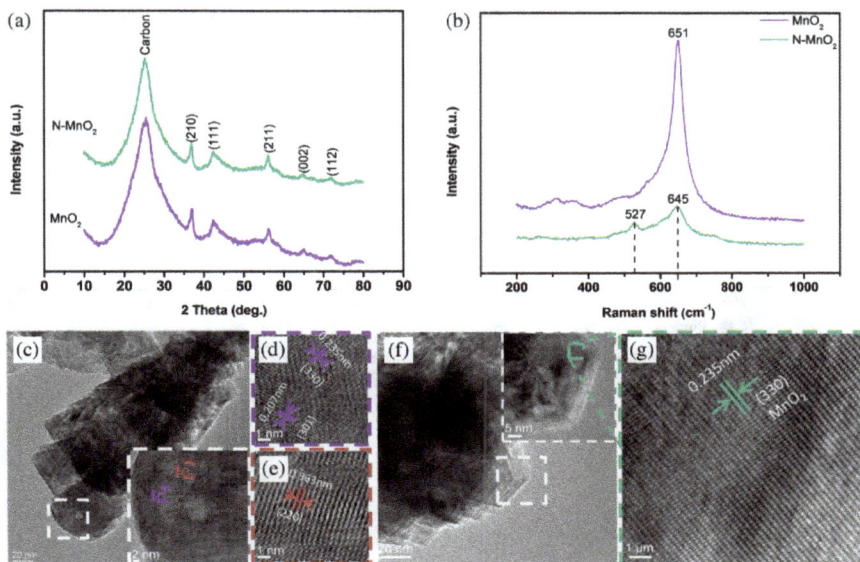

Fig. 2. (a) XRD profiles and (b) Raman spectra of MnO$_2$ and N-MnO$_2$ NRs on a carbon cloth. TEM and HRTEM of MnO$_2$ (c–e) and N-MnO$_2$ (f, g) NRs on a carbon cloth.

with the MnO$_2$ sample, demonstrating that the sample after N doping is still mainly MnO$_2$.[34,35]

Transmission electron microscopy (TEM) image shows that the MnO$_2$ sample is the nanorod structure (Fig. 2(c)). The high-resolution TEM (HRTEM) images (Figs. 2(c) and 2(d)) reveal clear lattice fringes, which are measured to be 0.207 nm, 0.235 nm and 0.363 nm, assigned to (301), (330) and (220) plane of β-MnO$_2$ (JCPDS # 24-0735), respectively. Figures 2(f) and 2(g) show the TEM and HRTEM of N-MnO$_2$, the measured lattice fringe of 0.235 nm matches well with the (330) plane of β-MnO$_2$ (JCPDS # 24-0735), confirming again that the heat-treated samples under NH$_3$ atmosphere did not undergo phase transition.

X-ray photoelectron spectroscopy (XPS) characterization was conducted as well to further study the composition of N-MnO$_2$ NRs. As shown in Fig. 3(a), in comparison to MnO$_2$ NRs sample, the N-MnO$_2$ exhibits distinct N signals and the percentage of N atoms in

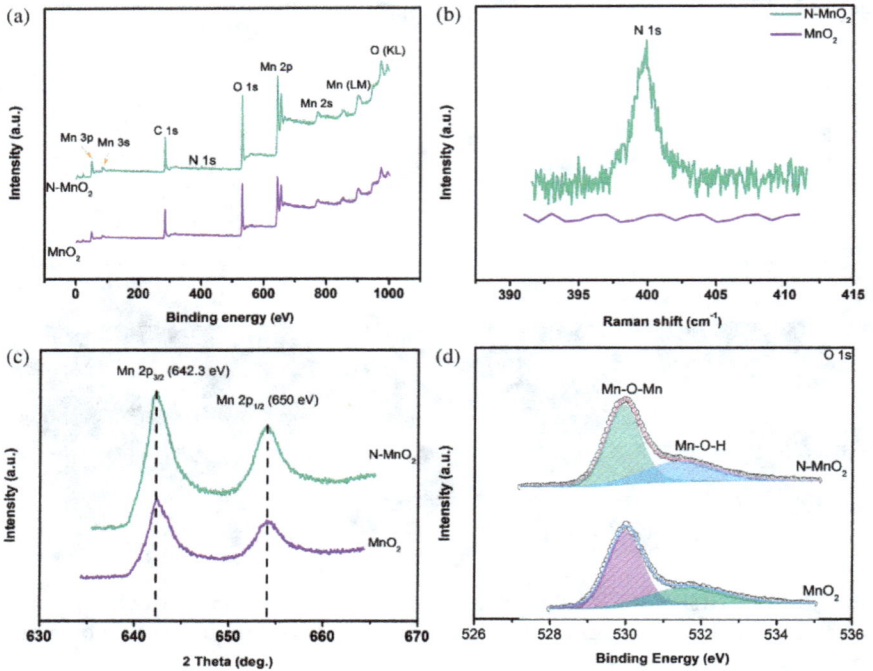

Fig. 3. (a) Survey XPS spectra, (b) N 1s, (c) Mn 2p and (d) O 1s core level XPS spectra for MnO_2 and $N-MnO_2$.

total is 2.33%, indicative of the doping of N element in MnO_2 NRs after annealing in NH_3 at 100°C. The N 1s core level spectra of both samples confirm again that $N-MnO_2$ sample possesses the element of N, while the MnO_2 sample does not (Fig. 3(b)). Furthermore, the Mn 2p characteristic peaks of both MnO_2 and $N-MnO_2$ samples were presented in Fig. 3(c), in which two peaks appearing at 642.3 eV and 650 eV can be indexed well to the typical Mn $2p_{3/2}$ and Mn $2p_{1/2}$ peaks of Mn^{4+}.[36,37] There is no obvious difference between the two spectra, suggesting the similar valence state of Mn in both samples. The O 1s peaks of $N-MnO_2$ are wider than that of MnO_2 (Fig. 3 (d)), revealing that the number of the valence bonds of Mn–O–Mn and Mn–O–H get increased during the thermal treatment in NH_3.[38]

To evaluate the effect of N-doping on the electrochemical properties of MnO_2 NRs and the feasibility of $N-MnO_2$ as the cathode material for zinc ion battery, two-electrode $Zn-MnO_2$ batteries were

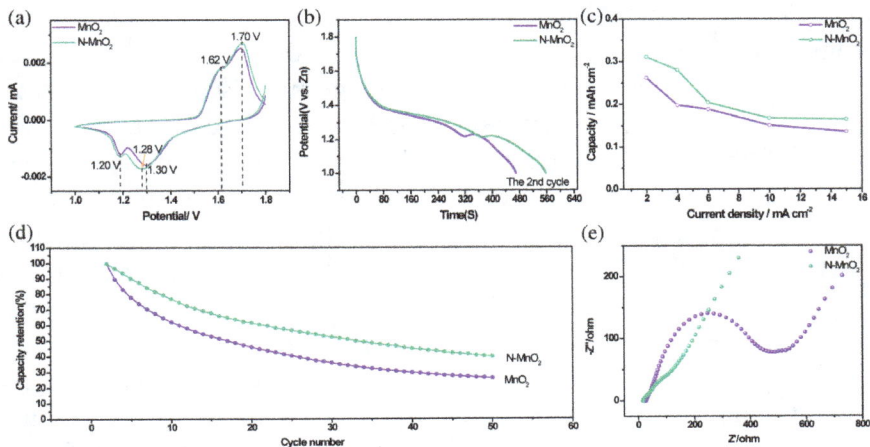

Fig. 4. (a) CV curves at 2 mV s^{-1}, (b) Galvanostatic discharge curves at 2 mA cm^{-2}, (c) specific capacity as a function of current density, (d) cycling performance at 10 mA cm^{-2} of the Zn-MnO$_2$ and Zn-N-MnO$_2$ battery and (e) Nyquist plots for Zn-MnO$_2$ and Zn-N-MnO$_2$ battery.

assembled using MnO$_2$ or N-MnO$_2$ as cathode (denoted as Zn-MnO$_2$ and Zn-N-MnO$_2$), Zn foil as anode in 2 M ZnSO$_4$ and 0.4 M MnSO$_4$ mixed aqueous electrolyte. Cyclic voltammograms (CVs) profiles of the Zn-MnO$_2$ and Zn-N-MnO$_2$ batteries are plotted in Fig. 4(a), where both batteries exhibit two pairs of redox peaks, oxidation peaks in 1.62 V, 1.70 V and reduction peaks in 1.20 V, 1.28 V and 1.30 V, suggesting the two step Faradic reaction. Figure 4(b) compares galvanostatic discharge curves of the Zn-MnO$_2$ and Zn-N-MnO$_2$ batteries. Both curves show two discharge plateaus, which is consistent with the CV results. The Zn-NMnO$_2$ battery presents higher discharge plateau and longer discharge time than that of Zn-MnO$_2$ battery with similar mass loading (2.58 mg cm^{-2} for MnO$_2$ and 2.56 mg cm^{-2} for N-MnO$_2$), indicating N-doping can improve the discharge capacity of MnO$_2$ NRs. Figure 4(c) plots the calculated capacities of the two batteries as a function of current densities. The Zn-N-MnO$_2$ battery achieves higher capacity at various current densities. Significantly, the Zn-N-MnO$_2$ battery yields the highest capacity of 0.31 mA h cm^{-2} at 2 mA cm^{-2}, while the Zn-MnO$_2$ battery only delivers 0.26 mAh cm^{-2} at the same current density. In addition, Zn-N-MnO$_2$ battery owns relatively higher durability than that of Zn-MnO$_2$ battery within

50 cycles (Fig. 4(d)). Nevertheless, the Zn-N-MnO$_2$ battery still suffers from a low cycling stability owing to the dissolution and irreversible reaction of MnO$_2$ cathode during the charge and discharge processes. The possible strategies, such as compounding with conductive polymers,[21] forming composites with carbon materials,[39] optimizing the structures of MnO$_2$,[40] could be taken to prevent the dissolution and reduce the accumulation of the byproducts during the charge and discharge processes. To further investigate the electrochemical property of Zn-MnO$_2$ and Zn-N-MnO$_2$ battery, electrochemical impedance measurements were also conducted. As shown in Fig. 4(e), the plots display a straight line in the low-frequency region and a semicircle in the high-frequency region, corresponding to ion diffusion limited processes and electron transfer limited processes, respectively. The charge transfer resistance (R_{ct}) of Zn-N-MnO$_2$ battery is ~16.4 Ω, much smaller than that of Zn-MnO$_2$ battery. Moreover, the linear slope of Zn-N-MnO$_2$ battery is higher than that of Zn-MnO$_2$ battery, demonstrating Zn-N-MnO$_2$ battery has smaller ion diffusion (Z_w) as well. These findings illustrate that the electronic conductivity and ion diffusion of MnO$_2$ electrode has been improved after N doping.

Figure 5 shows the Ragone plots (energy density vs. power density) by comparing the Zn-N-MnO$_2$ battery with reported γ-MnO$_2$ (160.4 Wh kg^{-1}

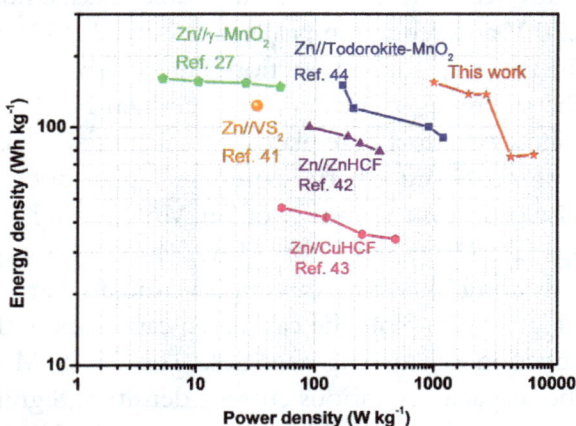

Fig. 5.   Ragone plot of the Zn-N-MnO$_2$ battery compared with some other known cathode materials for aqueous Zinc ion batteries.

Source: Data taken from Refs. 27 and 41–44.

at 5.21 W kg^{-1}),[27] VS$_2$ (123 Wh kg^{-1} at 32.32 W kg^{-1}),[41] ZnHCF (100 Wh kg^{-1} at 90 W kg^{-1}),[42] CuHCF (45.7 Wh kg^{-1} at 52.5 W kg^{-1}),[430] and Todorokite-MnO$_2$ (150 Wh kg^{-1} at 170 W kg^{-1})[44] cathodes for aqueous zinc ion batteries. High energy density and power density (154.3 Wh kg^{-1} at 1003.9 W kg^{-1}, and 76.4 Wh kg^{-1} at 6914.7 W kg^{-1}) can be achieved, which is promising for energy storage applications.

4. Conclusion

In summary, the conductivity and electrochemical property of MnO$_2$ have been intrinsically improved by introducing N into MnO$_2$ NRs. The Zn-N-MnO$_2$ battery based on the N-MnO$_2$ cathode displayed the highest capacity of 0.31 mAh cm^{-2} at 2 mA cm^{-2} with relatively better cyclic stability. In addition, the Zn-N-MnO$_2$ battery achieved a maximum energy density of 154.3 Wh kg^{-1} at a current density of 2 mA cm^{-2} and a highest power density of 6914.7 W kg^{-1} at 15 mA cm^{-2}. Thermal treatment under NH$_3$ atmosphere has proved to be an efficient way to induce N element into MnO$_2$ NRs, thereby improving the electrochemical performance of aqueous Zn-MnO$_2$ battery. This modification method is simple, safe and easy to operate, providing a promising direction for the future development of Zn-MnO$_2$ batteries.

Acknowledgments

This work was supported by the Guangdong Natural Science Funds for Distinguished Young Scholar (2014A030306048), Tip-top Scientific and Technical Innovative Youth Talents of Guangdong Special Support Program (2015TQ01C205), Pearl River Nova Program of Guangzhou (201610010080), Province key platforms and projects of Guangdong Universities (2017KQNCX195), and the Fundamental Research Funds for the Central Universities (17lgzd16).

References

1. S. Grugeon *et al.*, *Chem. Mater.* **17**, 5041 (2005).
2. L. Mai *et al.*, *Chem. Rev.* **114**, 11828 (2014).

3. D. Yu *et al.*, *Nat. Nanotechnol.* **9**, 555 (2014).
4. P. Simon *et al.*, *Science* **343**, 1210 (2014).
5. Q.-C. Liu *et al.*, *Nat. Commun* **6**, 7892 (2015).
6. J. Nawishta *et al.*, *Adv. Mater.* **29**, 1700804 (2017).
7. Y. Zeng *et al.*, *Adv. Mater.* **30**, 1707290 (2018).
8. P. Zhang *et al.*, *J. Mater. Chem. A* **6**, 8895 (2018).
9. K. Xu *et al.*, *Funct. Mater. Lett.* **10**, 1750025 (2017).
10. J. Ni *et al.*, *Funct. Mater. Lett.* **9**, 1650004 (2016).
11. M. S. Whittingham, *Chem. Rev.* **104**, 4271 (2004).
12. J. B. Goodenough *et al.*, *Chem. Mater.* **22**, 587 (2009).
13. J.-M. Tarascon *et al.*, *Materials For Sustainable Energy: A Collection of Peer-Reviewed Research and Review Articles from Nature Publishing Group* (World Scientific, 2011), pp. 171–179.
14. P. G. Bruce *et al.*, *Nat. Mater.* **11**, 19 (2012).
15. X. Dong *et al.*, *Angew. Chem. Int. Ed.* **55**, 7474 (2016).
16. F. Beck *et al.*, *Electrochim. Acta* **45**, 2467 (2000).
17. J. Yan *et al.*, *J. Power Sources* **216**, 222 (2012).
18. Y. Zeng *et al.*, *Adv. Mater.* **30**, 1802396 (2018).
19. X. Cheng *et al.*, *Electroim. Acta* **263**, 311 (2018).
20. W. Sun *et al.*, *J. Am. Chem. Soc.* **139**, 9775 (2017).
21. Y. Zeng *et al.*, *Adv. Mater.* **29**, 1700274 (2017).
22. M. H. Alfaruqi *et al.*, *Chem. Mater.* **27**, 3609 (2015).
23. X. Yu *et al.*, *Nano Energy* **2**, 1242 (2013).
24. N. Zhang *et al.*, *Nat. Commun.* **8**, 405 (2017).
25. Z. Wang *et al.*, *Adv. Mater.* **26**, 970 (2014).
26. C.-C. Yang *et al.*, *J. Power Sources* **112**, 174 (2002).
27. F. Cheng *et al.*, *Adv. Mater.* **17**, 2753 (2005).
28. C. Xu *et al.*, *Angew. Chem. Int. Ed.* **51**, 933 (2012).
29. H. Pan *et al.*, *Nat. Energy* **1**, 16039 (2016).
30. M. H. Alfaruqi *et al.*, *Electrochem. Commun.* **60**, 121 (2015).
31. J. K. Seo *et al.*, *J. Phys. Chem. C.* **122**, 11177 (2018).
32. H. Xia *et al.*, *Funct. Mater. Lett.* **2**, 13 (2009).
33. H. Li *et al.*, *Electrochim. Acta* **164**, 252 (2015).
34. D. Gosztola *et al.*, *J. Electroanal. Chem. Interf. Electrochem.* **271**, 141 (1989).
35. E. Widjaja *et al.*, *Anal. Chimi. Acta* **585**, 241 (2007).
36. F. Xiao *et al.*, *Int. J. Electrochem.* Sci. **7**, 7440 (2012).
37. C.-H. Liang *et al.*, *J. Alloy. Compd.* **500**, 102 (2010).

38. R. S. Kalubarme *et al.*, *Electrochim. Acta* **87**, 457 (2013).
39. B. Wu *et al.*, *Small* **14**, 1703850 (2018).
40. H.-W. Zhu *et al.*, *Nano Res.* **11**, 1554 (2018).
41. P. He *et al.*, *Adv. Energy Mater.* 7, 1601920 (2017).
42. L. Zhang *et al.*, *Adv. Energy Mater.* **5**, 1400930 (2015).
43. R. Trócoli *et al.*, *ChemSusChem* **8**, 481 (2015).
44. J. Lee *et al.*, *Electrochim. Acta* **112**, 138 (2013).

Chapter 13

Metal-Organic Framework–Derived Structures for Next-Generation Rechargeable Batteries

Wenhui Shi[*,‡], Xilian Xu[†], Lin Zhang[†], Wenxian Liu[†], and Xiehong Cao[†,§,¶]

Metal-organic frameworks (MOFs) have attracted great attention as versatile precursors or sacrificial templates for the preparation of novel porous structures. Due to their tunable compositions, structures and porosities as well as high surface area, MOF-derived materials have revealed promising performance for energy storage devices. In this mini review, the recent progress of MOF-derived materials as electrodes of next-generation rechargeable batteries was summarized. We briefly introduce the preparation methods, various design strategies and the structure-dependent performance of recently reported MOF-derived materials as electrodes of post-lithium-ion batteries, focusing on lithium-sulfur (Li-S) batteries, sodium-ion batteries

[*]Center for Membrane and Water Science & Technology, Ocean College, Zhejiang University of Technology, Hangzhou 310014, P. R. China.

[†]College of Materials Science and Engineering, Zhejiang University of Technology, Hangzhou 310014, P. R. China.

[‡]Huzhou Institute of Collaborative Innovation Center for Membrane Separation and Water Treatment, Zhejiang University of Technology, Huzhou, Zhejiang 313000, P. R. China.

[§]Institute of Advanced Electrochemical Energy, Xi'an University of Technology, Xi'an 710048, P. R. China.

[¶]gcscaoxh@zjut.edu.cn

(SIBs) and metal–air batteries. Finally, we give the conclusion with some insights into future development of MOF-derived materials for next-generation rechargeable batteries.

Keywords: Metal-organic frameworks; MOF-derived materials; lithium-sulfur batteries; sodium-ion batteries; metal–air batteries.

1. Introduction

Due to vast use of fossil fuels, development of highly effective, renewable and clean resources of energy is becoming a global concern. Many efforts have been made for the development of various devices applicable for the energy storage and conversion.[1-5] Among them, Lithium-ion batteries (LIBs) are in an extensive use for rechargeable batteries that possess high energy density, high energy efficiency and long lifespan. Nevertheless, high cost, comparatively lower power density along with safety issues restrict their largescale applications in the portable devices and electric vehicles (EVs).[6-11] Therefore, it requires finding some alternative next-generation rechargeable batteries that possess high power and energy density. Recently, new types of rechargeable batteries have been introduced, such as lithium-sulfur (Li-S) batteries,[12,13] sodium-ion batteries (SIBs)[14-16] and metal–air batteries,[17-20] which are emerging as promising candidates for post-lithium-ion batteries. It is well known that the properties of electrode materials of rechargeable batteries are considered as one of the most important factors determining the performance. Therefore, developing advanced electrode materials that have high specific capacity, excellent rate capability and long cycle life is a critical issue for next-generation rechargeable batteries.[21-26]

Metal-organic frameworks (MOFs), a large family of porous crystals, are formed through linkage of metallic ions or clusters with organic ligands. One of the unique advantages of MOFs is that their structure and porosity are designable through selection of certain metal species and organic linkers.[27-30] Nevertheless, the lower electrical conduction and stability of MOFs have restricted their usage. Recently, MOFs have attracted great attention as versatile templates to derive novel porous structures, for example, porous carbon,[31-34] metal oxides[35,36] or other metal compounds (i.e. metal sulfides, phosphides, selenides or carbides)[37-42] and multifunctional hybrids.[43-45] Owing to their tunable compositions, structures, and porosities as well as high surface areas, MOF-derived materials have revealed promising performance for energy storage devices.

Generally, MOF-derived materials are prepared via a simple thermolysis process at an elevated temperature under a certain atmosphere condition (for example, argon, air or nitrogen). During this process, the decomposition of organic ligands takes place to form carbon, and metal species are transformed into metal and/or metal oxide which will evaporate leaving pores at high temperatures or act as catalysts to induce the formation of graphitized carbons.[46-48] Furthermore, various heteroatoms such as boron, sulfur or nitrogen are able to be doped in the MOF-derived materials through use of certain organic ligands, for example, methyl imidazole as nitrogen-containing ligand, which is beneficial to tune the electronic feature and surface wettability.[15,49-52] Until now, MOF-derived materials that possess different morphologies and microstructures have been developed, from zero-dimensional (0D) polyhedrons[53-55] to one-dimensional (1D) rods,[56-60] two-dimensional (2D) nanosheets[61-63] and three-dimensional (3D) networks or arrays.[64-67] Additionally, MOFs are able to hybrid with other functional materials for constructing hierarchical structures.[68-70] Therefore, nanomaterials with various compositions and structures can be obtained using MOFs as precursors, which provide highly promising strategy to solve the key challenges of next-generation rechargeable batteries (Fig. 1).

In this mini review, we will focus on the recent research progress in the preparation of MOF-derived materials and their promising applications in next-generation rechargeable batteries. We will briefly introduce the synthetic strategies of different MOF-derived materials followed by discussion of their performance as electrodes of post-lithium-ion batteries, including Li-S batteries, SIBs and metal–air batteries. Finally, the conclusion with some future outlooks in this research field are presented.

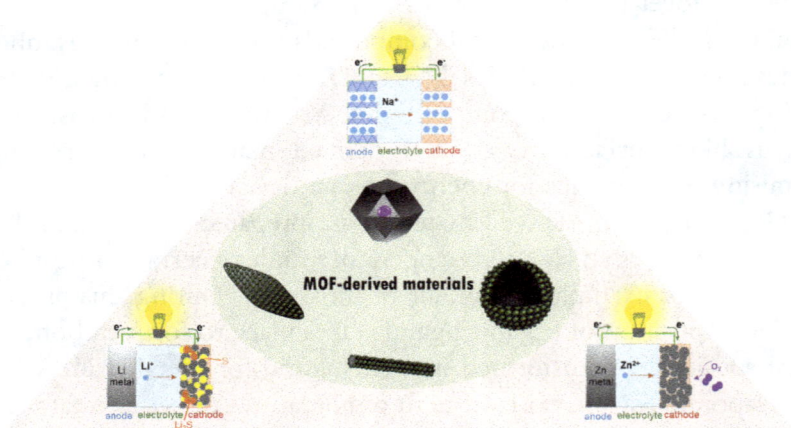

Fig. 1. Schematic illustration of different MOF-derived materials applied for next-generation rechargeable batteries.

2. MOF-Derived Materials for Next-Generation Rechargeable Batteries

2.1. *Lithium-sulfur batteries*

As a promising rechargeable battery, lithium-sulfur (Li-S) batteries have shown extraordinary potential with a high theoretical specific capacity of 1672 mAh g^{-1} and an energy density of 2600 Whk g^{-1} by using abundant and environmentally friendly sulfur as cathode material.[71-74] However, several problems need to be addressed for Li-S batteries. Using electrically insulating sulfur (S$_8$) as cathode, Li-S batteries suffer from low active material utilization and poor rate capability. Moreover, lithium polysulfides Li$_2$S$_n$ ($4 \leq n \leq 8$) formed as an intermediate product are solvable in electrolyte. The dissolved polysulfides may cause the "shuttle effect" and react with the lithium anode, resulting in a rapid decline of capacity and a low coulombic efficiency.[75] In order to overcome this problem, one of the most efficient ways is encapsulation of sulfur into carbon host to construct a carbon/sulfur (C–S) hybrid. Novel carbon materials that possess large surface area, sufficient pores and hierarchical structures have been considered as excellent sulfur host as cathode material owing to their ability to effectively improve electrical conductivity, accommodate the huge volume change as well as firmly immobilize sulfur.

Recently, porous carbon materials derived from MOFs are emerging as favorable candidates used in the Li-S batteries owing to large surface area, abundant micropores, good electrical conductivity, etc. Xu *et al.*[76] first encapsulated sulfur in the porous carbon nanoplates obtained by one-step pyrolysis of MOF-5. The obtained C–S composite cathode displayed good cycling performance and a high specific capacity of 730 mAh g^{-1} after 50 cycles charge/discharge at a current density of 837.5 mA g^{-1}. Later on, Wu *et al.*[74] reported ZIF-8 derived microporous carbon polyhedrons as carbon host to form C-S composites. Well-prepared carbonic polyhedrons that have sufficient and unvarying micropores were later on used as carbon host for preparation of carbon/sulfur mixtures. Their research showed that the temperature for sulfur loading, sulfur content and the electrolyte affected the performance of Li-S batteries. Although enhancement of the reversible capacity and cycling stability have been achieved by early studies on the MOF-derived porous C-S composite cathode for Li–S batteries, the overall performance is still not satisfactory for commercial application.

To further eliminate the loss of active sulfur, Li *et al.* for the first time encapsulated small sulfur molecules of S_{2-4} within MOF-derived microporous nitrogen-doped carbon (Fig. 2).[77] All of the small S_{2-4} molecules were confined in the micropores with a size of 0.5 nm, which enhanced the utilization of sulfur and reduced the dissolution of formed polysulfides. The nitrogen-doped carbon/S_{2-4} hybrid cathode delivered a specific capacity as high as 936.5 mAh g^{-1} after 100 cycles charge/discharge at a current density of 335 mA g^{-1}. Moreover, it also showed outstanding rate capability, which exhibited a specific capacity of 632 mAh g^{-1} at a high current density of 5Ag^{-1}. The superior performance of the above nitrogen-doped carbon/S_{2-4} hybrid electrode can be attributed to the firm host of small S_{2-4} molecules in the micropores along with relief of mechanical stress through the hierarchical structure of MOF-derived carbon with both micropores and mesopores. In addition, N-doping in MOF-derived carbon can facilitate the fast charge transfer. Meanwhile, it can create a strong interaction between carbon and sulfur, thus firmly immobilizing

(a)

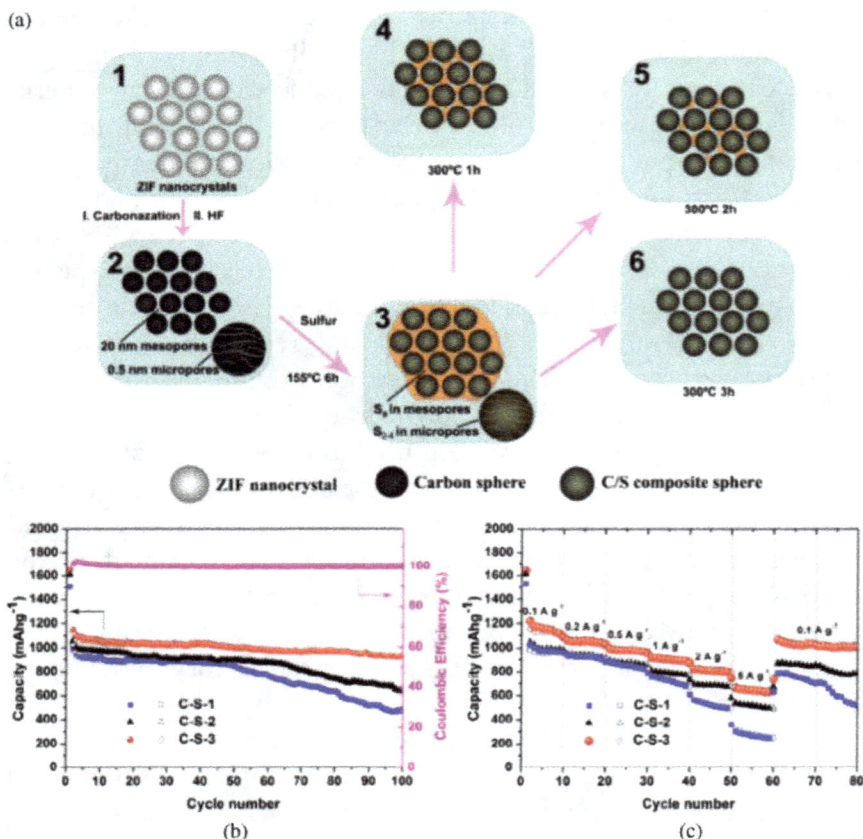

Fig. 2. (a) Schematic illustration for the preparation process of C–S hybrids. (b) Cycling stability and Coulombic efficiency of C–S hybrids at a current density of 335 mA g^{-1}. (c) Rate capability of C–S hybrids. Reproduced with permission from Ref. 77. Copyright 2015, American Chemical Society.

sulfur in micropores. Similar strategies have also been adopted by other works. For example, Hong *et al.* developed a flower like N-doped microporous carbon assembled by nanosheets, which also acted as an ideal carbon host of small S$_{2-4}$ molecules.[73]

Porous carbon materials with varied morphologies can be derived from MOFs and used as efficient sulfur immobilizer, such as 2D nanosheets,[78] carbon nanocages[46] and so on. More importantly, carbon/sulfur composites derived from MOFs with more complicated structures and compositions have been developed and used for Li-S batteries recently.[79,80] For instance, cobalt-doped porous carbon poly-hedrons were synthesized by using ZIF-67 polyhedrons as the precursor and then wrapped with graphene oxide (GO) nanosheets[47] After the reduction process of GO to reduced graphene oxide (rGO) and the subsequent sulfur impregnation, the rGO/C–Co–S polyhedron hybrid was obtained (Fig. 3). The rGO/C–Co–S electrode exhibited great improvement in electrochemical performance. After 300 cycles, it delivered a high specific capacity of 949 mAh g^{-1} at a current density

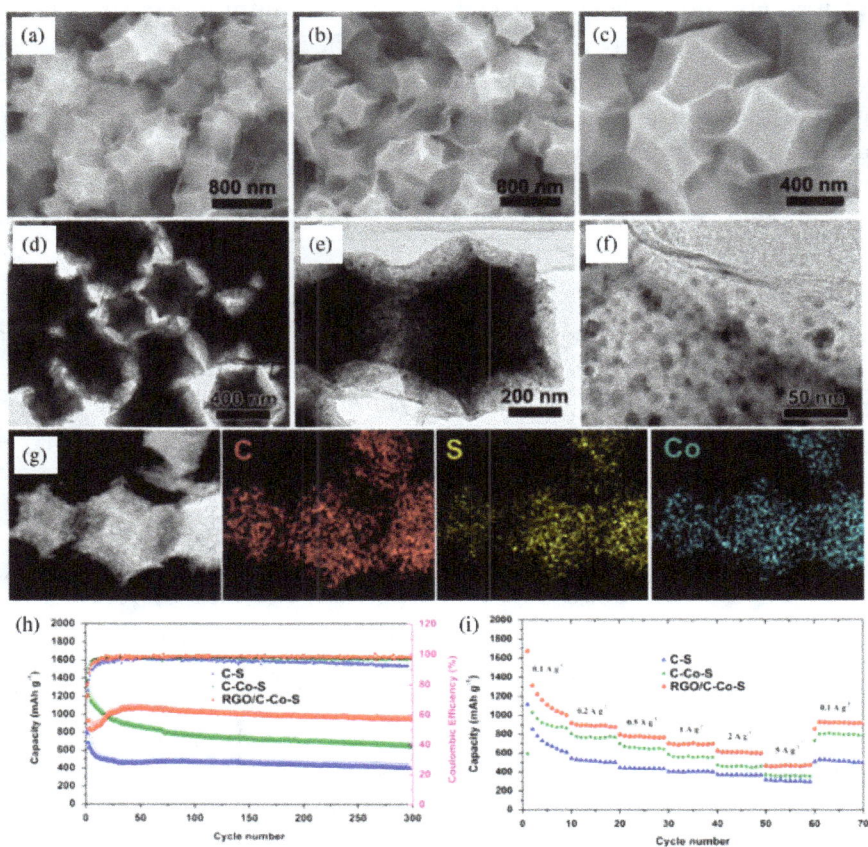

Fig. 3. (a–c) SEM images, (d–f) TEM images and (g) Elemental mapping of rGO/C–Co–S polyhedrons. (h) Cycling stability and Coulombic efficiency of rGO/C–Co–S polyhedron cathode at a current density of 0.3 Ag^{-1} for 300 cycles. (i) Rate capabilities at different current densities. Reproduced with permission from Ref. 47. Copyright 2016, Elsevier Ltd.

of 0.3 A g^{-1}. The rGO/C–Co–S cathode also displayed excellent rate capability, delivered the capacities of 772, 704 and 606 mAh g^{-1} at current densities of 0.5, 1 and 2 A g^{-1}, respectively. The high reversible capacity and rate capability were attributed to the unique structures, in which the carbon host with sufficient pores physically adsorbs sulfur molecules, and the small Co particles form strong interactions with sulfur. Furthermore, the rGO nanosheets act as an additional barrier to further reduce the dissolution of polysulfides. Therefore, the porous carbon, Co nanoparticles and rGO nanosheets have synergistic effects which work together to firmly immobilize the sulfur molecules and greatly relieve the shuttle effect.

Based on above research works, MOF-derived porous carbon with properly designed structure and compositions can be promising candidate for sulfur host as the cathode of Li–S batteries (Table 1).[47,81]

Table 1. Summary of MOF-derived materials for Li–S batteries.

Original MOF	Sample	Capacitance (discharge capacity (mAh g^{-1})/rate mA (mA g^{-1}))	Cycling stability (reversible capacity (mAh g^{-1})/rate mA (mA g^{-1})/cycling number)	Ref.
ZIF-67	RGO/C–Co–S	1218/300	949/300/300	47
ZIF-67	GC–Co–S	1255/0.2C	992/0.2C/100	46
ZIF-67	N–Co$_3$O$_4$@N–C/rGO	1223/0.2C	611/2C/1000	68
ZIF-67	S@Co–N–GC	1150/1C	625/1C/500	96
ZIF-8	S/ZIF-8	—	553/0.5C/300	98
ZIF-8	NDS–sulphur	1655.7/335	936.5/335/100	77
ZIF-8 nanosheet	S/ZIF-8-NS-C	1226/0.2 C	587/0.5C/300	78
Cu-TDPAT	S@Cu-TDPAT (100 nm)	995/0.5C	745/1C/500	72
Zn-TDPAT	FMNCN-900	1645/0.1C	1220/0.1C/200	73
PCN-224	ppy-S-in-PCN-224	—	670/10.0C/200	13
NH$_2$-MIL-101(Al)	S/CN-5@NSHPC	1099/0.1C	445/1C/500	52
HKUST-1	S@HKUST-1/CNT	1263/0.2C	757/0.2C/500	99
Na$_2$Fe[Fe(CN)$_6$]	S@Na$_2$Fe[Fe(CN)$_6$]@PEDOT	1291/0.1C	770/1C/100	97

Generally, the advantages of MOF-derived porous carbon for Li–S battery applications include the following: (1) high porosity enables a high sulfur loading; (2) narrow distribution of micropores can effectively trap and immobilize sulfur and thus reduce the dissolution of polysulfides; (3) hierarchically porous structures can induce the fast transport of lithium ions leading to the excellent rate performance; and (4) heteroatom or transition metal doping of porous carbon can enhance the electrical conductivity. Meanwhile, it can improve the interaction between porous carbon and sulfur.

2.2. Sodium-ion batteries

Recently, SIBs have gained great attention as a promising alternative to LIBs due to the abundance of sodium, low cost, and appropriate redox potential. Nevertheless, as compared to lithium, insertion/extraction of sodium ions lead to much larger volume change in the electrode due to its larger ionic radius.[82–84] Therefore, exploring high-performance electrode materials is of great importance for the advancement of SIBs. A number of cathode materials for SIBs have been developed, such as $NaFePO_4$,[85–87] $Na_3V_2(PO4)_3$,[88–90] and $NaMnO_2$.[91–93] Up to now, only a few materials are available for anode of SIBs[94,95] Therefore, there is an urgent need to find novel anode materials for SIBs achieving both high capacity and long cycle life.

MOFs as versatile precursors to produce various porous nanomaterials, including porous carbons, metal oxides, metal sulfides, metal phosphides, and metal carbides, as well as new multifunctional hybrid materials, have demonstrated to be excellent electrodes of SIBs (Table 2).[31,38,100,101] For example, nitrogen-doped carbon-coated Co_3O_4 nanoparticles were derived from ZIF-67. When explored as anode of SIBs, it delivered a specific capacity of 373 mAh g^{-1} after 60 cycles charge/discharge at a current density of 200 mA g^{-1}. Similarly, Fang *et al.* reported 2D bimetallic oxide (Co_3O_4/ZnO) nanosheets using bimetallic MOF as the precursor.[102] The Co_3O_4/ZnO hybrid nanosheets with rich oxygen vacancies have abundant porosity and high surface areas. When used as anode of SIBs, the Co_3O_4/ZnO hybrid nanosheets exhibited high rate capability with a specific

Table 2. Summary of MOF-derived materials for SIBs.

Original MOF	Sample	Capacitance (discharge capacity (mAh g^{-1})/rate mA (mA g^{-1}))	Cycling stability (reversible capacity (mAh g^{-1})/rate mA (mA g^{-1})/ cycling number)	Ref.
ZIF-67	CoSe/C composites	—	531.6/500/50	38
ZIF-67	Co_3O_4@NC	813/100	175/1000/1100	105
ZIF-67	CoP@C-RGO-NF	1163.5/100	473.1/100/100	106
Co-MOF	7-CoS/C	—	542/1000/2000	101
ZIF-8	NPC	677.2/50	144.7/50/200	31
ZIF-8	$ZnS-Sb_2S_3$@C core-double shell	1675/100	630/100/120	104
ZIF-8	P@N-MCP	1312/150	600/150/100	15
MOF-5	CPC	1822.8/100	240/100/100	32
CoZn-MOF	$CoZn-O_2$	—	242/2000/1000	61
MIL-125(Ti)	TiO_2@C	393.3/100	148/500/500	84
MIL-125	Porous cake-like TiO_2	—	250/50/50	107
Ti-MOF	CRT	324.7/0.5C	175/0.5C/200	108
PB	RGO@CoP@FeP	968.0/100	456.2/100/200	39
PB	Porous $CoFe_2O_4$ nanocubes	573/50	360/50/50	109
Fe-MOF	FTO⊂CNT	965/100	358.8/100/200	110
MIL-101(Fe)	hollow hybrid Fe_2O_3@ MIL-101(Fe)/C	1052/200	662/200/200	83
MOF $Fe_4(Fe(CN)_6)_3$	hollow $MgFe_2O_4$ microboxes	406/50	135/50/150	82
Ni-MOFs	NiO/Ni/Graphene composites	992/200	385/200/190	55
Cu-based MOFs	porous CuO/rGO	—	466.6/100/50	69
Cu-MOFs	CHO-1	—	415/50/50	54
Mn-BTC MOFs	Porous MnO@C nanorods	—	260/50/100	112
Sn-MOF	SnO_2 nanosphere	1030/50	417/50/100	100

capacity of 242 mAh g^{-1} at a current density as high as 2000 mA g^{-1}. In addition, it delivered a capacity retention of 91% after 1000 cycles charge/discharge. The enhanced performance is believed to be due to its 2D structure and synergistic effect of the hybrid materials. Besides, other metal compounds with various structures have been successfully synthesized from MOF precursors and used as anode of NIBs, such as ZnS nanoparticles decorated on nitrogen-doped porous carbon polyhedra,[37] CoS nanoparticles embedded in porous carbon nanorods,[101] core-shell porous FeP@CoP phosphide micocubes,[39] Urchin-like CoSe$_2$ assembled by nanorods[103] and carbon-encapsulated metal selenides.[24]

MOF-derived materials with hierarchical porous structure have been developed for SIBs recently. Particularly, constructing core-shell structured electrodes is helpful to accommodate the volume expansion generated during repeated charge-discharge, alleviate pulverization problem and effectively improve the charge transfer kinetics. For example, Dong *et al.* designed a zinc sulfide-antimony sulfide@carbon (ZnS–Sb$_2$S$_3$@C) core-doubled shell by using ZIF-8 as sacrificial template (Fig. 4).[104] First, in order to protect the morphology of ZIF-8, a resorcinol-formaldehyde (RF) layer was grown on the surface of ZIF-8 before fabricating the hollow ZnS@RF polyhedron by the sulfurization process.

During the sulfurization, ions were able to penetrate through the RF layer and react with the inner ZIF-8. Then, a simple metal cation exchange process between Zn^{2+} and Sb^{3+} was conducted, followed by an annealing process to produce the final ZnS-Sb$_2$S$_3$@C core-double shell polyhedron. The ZnS-Sb$_2$S$_3$@C core-double shell polyhedron was composed of ZnS inner core and a Sb$_2$S$_3$/C double shell. This unique structure is in favor of the easy electrolyte infiltration, leading to the reduction of the ion diffusion length and thus the enhancement of the reaction kinetics. The hierarchical porous structures would also alleviate the pulverization caused by sodium ion insertion/extraction. Furthermore, the carbon shell can improve the electric conduction, and serve as an additional protective layer to accommodate the volume extension. As expected, with such a hierarchical structure, ZnS–Sb$_2$S$_3$@C electrode illustrated superior

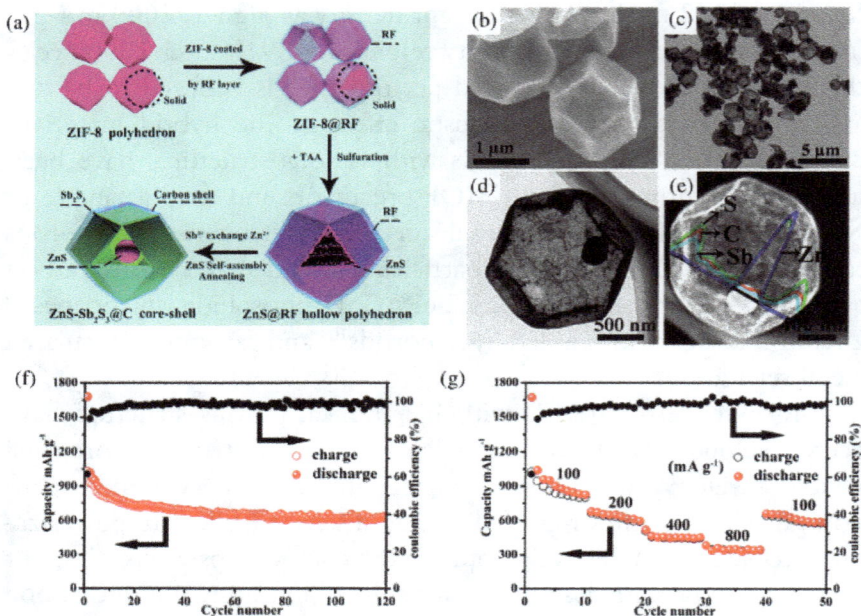

Fig. 4. (a) Scheme for the preparation process of ZnS–Sb$_2$S$_3$@C coredouble shell polyhedron composite. (b) FESEM image, (c, d) TEM images, and (e) Line scanning curves of ZnS–Sb$_2$S$_3$@C core-double shell polyhedron. (f) Cycling stability and (g) Rate performance at a current density of 100 mA g^{-1} of ZnS–Sb$_2$S$_3$@C core–shell as anode of SIBs. Reproduced with permission from Ref. 104. Copyright 2017, American Chemical Society.

electrochemical performance. A high specific capacity of 630 mAh g^{-1} was achieved at a rate of 100 mA g^{-1} after 120 cycles. This strategy of synthesizing MOF-derived metal sulfides can be extended to other nanostructured electrodes for high-performance rechargeable batteries.

Additionally, hierarchical structured electrode materials can be prepared by coating of MOFs on conductive substrates followed by annealing process. For example, core-shell structured CoP@C polyhedrons attached on 3D rGO sheets which are supported on nickel foam (NF) were synthesized through an *in-situ* phosphatization at low temperature (Fig. 5).[106,111] The anchoring of CoP homogeneously on rGO sheets can buffer the large volume change during

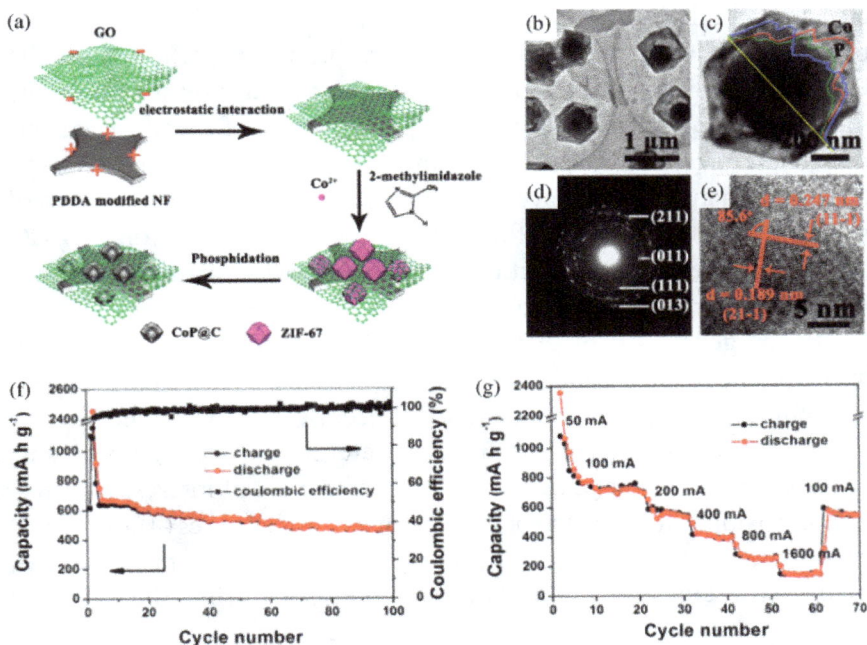

Fig. 5. (a) Scheme for the synthetic process of CoP@C–RGO–NF. (b, c) TEM images of CoP@C-RGO. (d) Electron diffraction pattern and (e) HRTEM image of single CoP nanoparticle. (f) Cycling stability and coulombic efficiency at a current density of 100 mA g^{-1} and (g) Rate capability of CoP@C–RGO–NF anode. Reproduced with permission from Ref. 106. Copyright 2016, Elsevier Ltd.

repeating charge and discharge and prevent the agglomeration of active materials. The interconnected networks of grapheme can effectively improve the charge transfer kinetics. Therefore, this unique CoP@C–RGO–NF binder-free electrode of SIBs showed an excellent performance of long cycling life and excellent rate capability. Specifically, it delivered a reversible capacity of 473.1 mAh g^{-1} after 100 charge/discharge cycles at a rate of 100 mA g^{-1}.

2.3. *Metal–air batteries*

Metal–air batteries are one of the most promising alternatives for post-lithium-ion batteries because of their high energy density, environmental friendliness and low cost.[115–117] Typical metal–air batteries consist of a metal anode, a separator, an air cathode and the electrolyte. Among these components, the air cathode is one of the most critical parts to determine the performance of metal–air batteries. Therefore, as the two most important reactions for metal–air batteries, many efforts have been dedicated to the development of nonnoble electrocatalysts for oxygen reduction reaction (ORR) and oxygen evolution reaction (OER) in order to achieve high-performance rechargeable metal–air batteries.[118–121]

Rechargeable Zn–air batteries have been extensively explored to improve the rechargeability and energy efficiency by employing an efficient catalyst on the air cathode. Recently, MOF-derived carbon materials with advantages of uniform heteroatom doping, large surface area and long-range order have attracted a lot of interests as alternatives to noble-metal electrocatalysts for Zn–air batteries (Table 3).[17,41] However, the above materials usually suffer from several problems, such as the collapse of the carbon framework during the pyrolysis

Table 3.   Summary of MOF-derived materials for Zn–air batteries.

Original MOF	Sample	Charge/discharge voltage (V)	Power density/capacity	Ref.
ZIF-67	3D-CNTA	1.99 V/1.31 V	157.3 mW cm^{-2}	113
ZIF-8	Mo–N/C@MoS$_2$	1.99 V/1.24 V/5 mA cm^{-2}	196.4 mW cm^{-2}/5 mA cm^{-2}	18
ZIF-8	3D ordered N-rich carbon photonic crystals	—	197 mW cm^{-2}/319 mA cm^{-2}	25
Cu-ZIF-8	Cu-N/C	—	132 mW cm^{-2}/204 mA cm^{-2}	41
MIL-53	Fe-MOF@CNTs-G	Initial voltage gap of 1.01 V	95.3 mW cm^{-2}/120 mA cm^{-2}	45
MC-BIF-1S	BNPC	2.19 V/1.16 V/2 mA cm^{-2}	—	114
ZnO@Zn/Co-ZIF nanowires	Co@CoO$_x$/NCNTs	—	353.3 mW cm^{-2}/541.7 mA cm^{-2}	123
[Co$_6$(MIDPPA)$_3$ (1,2,4-btc)$_3$(NO$_2$)$_3$ (H$_2$O)$_3$](H$_2$O)$_7$	C-MOF-C$_2$-900	1.81 V/1.28 V/2 mA cm^{-2}	105 mW cm^{-2}/5 mA cm^{-2}	17

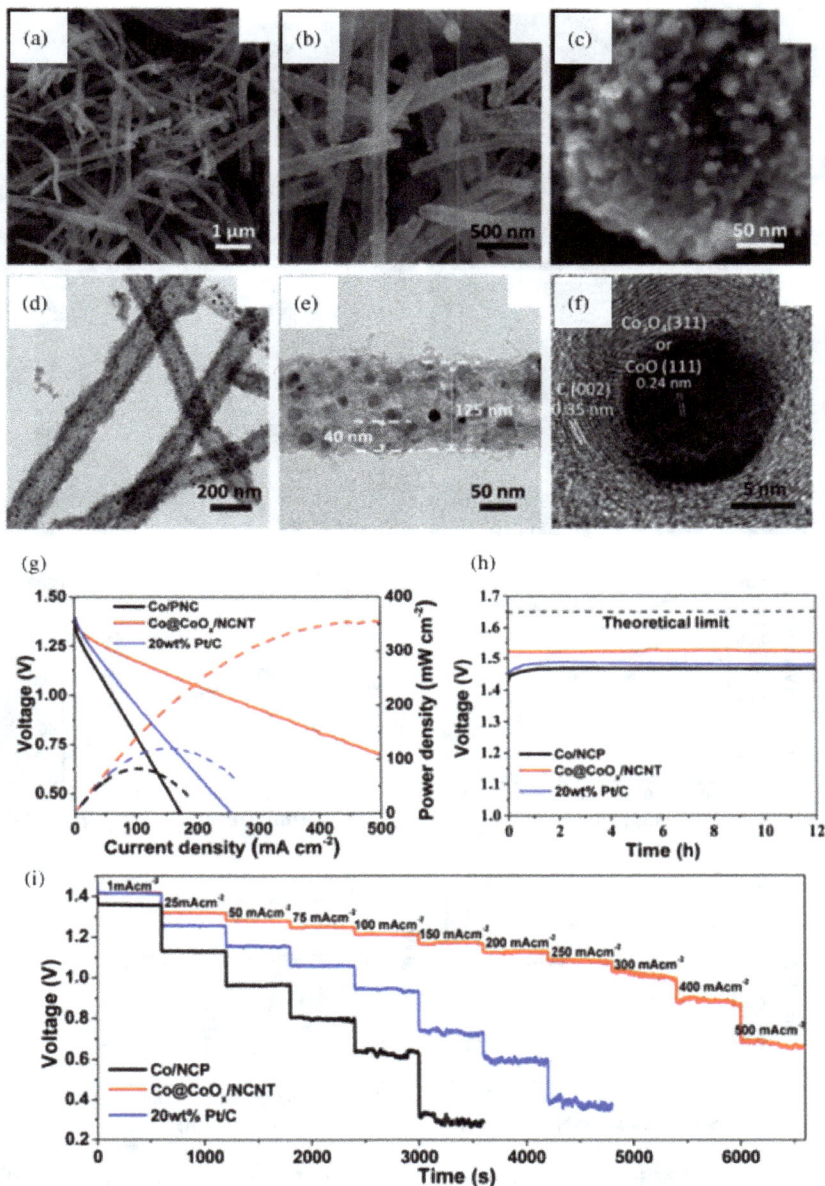

Fig. 6. (a, b, c) SEM images, (d, e) TEM and (f) HR-TEM images of Co@CoO$_x$/ NCNT. (g) Discharge current–voltage curve and power density vs. current density curve. (h) OCV vs. time curves. (i) Discharge curves at current densities ranging from 1 to 500 mA cm^{-2}. Reproduced with permission from Ref. 123. Copyright 2017, Royal Society of Chemistry.

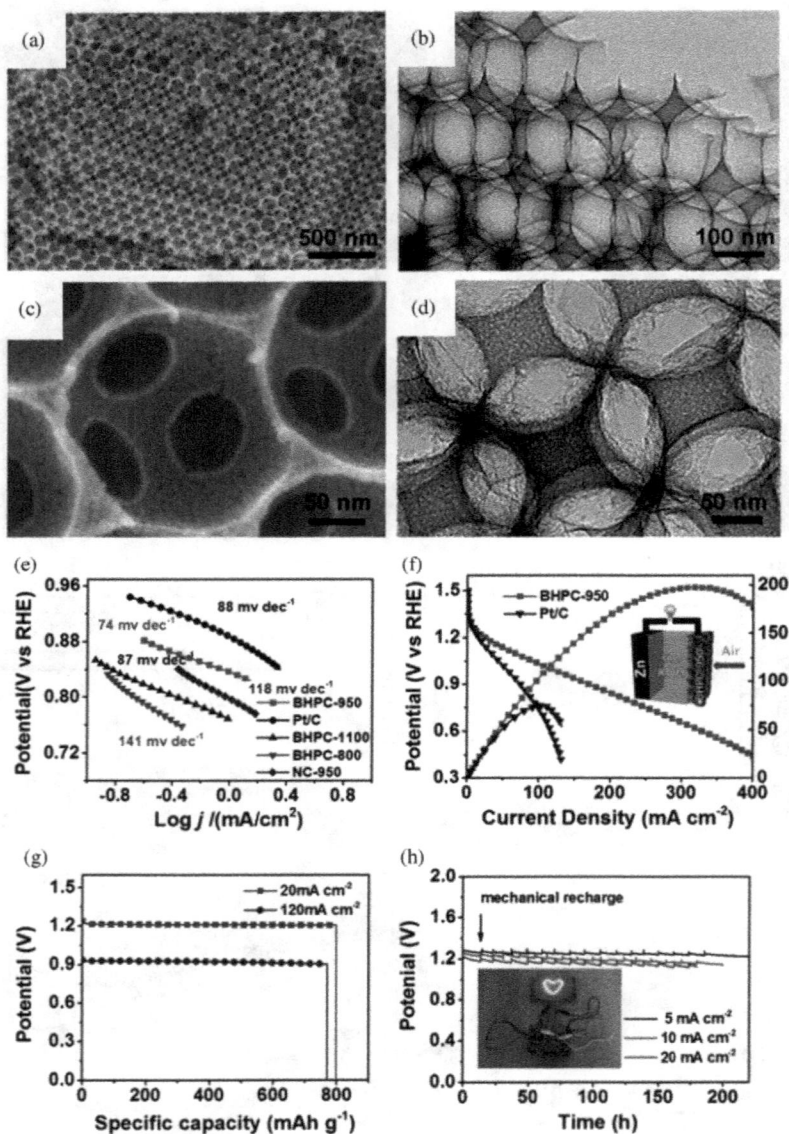

Fig. 7. (a–c) SEM images and (d) TEM image of BHPC-950. (e) Tafel plots for the ORR process of various electrodes. (f) Polarization and power density curves of Zn–air batteries with 20 wt% Pt/C and BHPC-950 as the cathode, respectively. (g) Specific capacities of BHPC-950-based Zn–air battery at current densities of 20 and 120 mA cm⁻². (h) Discharge curves at various current densities, the inset is the photo of a LED pattern powered by two BHPC-950 catalyzed Zn–air batteries connected in series. Reproduced with permission from Ref. 124. Copyright 2017, John Wiley and Sons.

process, narrow pore size distribution and thus insufficient exposure of catalytic active sites. Therefore, designing morphology and structure of MOF-derived electrocatalysts is of great importance.[122]

1D materials may achieve fast ORR kinetics at high discharge rate for metal–air batteries due to faster electron transportation along the 1D direction as well as shorter ion diffusion distance.[13] Lin *et al.* synthesized a 1D bimetallic MOF by using ZnO nanowires as the template and after the following calcination process, N-doped carbon nanotubes which were dispersed with the core–shell Co@CoO$_x$ nanoparticles (Co@CoO$_x$/NCNTs) were obtained (Fig. 6).[123] Based on Co@CoO$_x$/NCNTs, the assembled Zn–air batteries showed high open-circuit voltage of 1.52 V, excellent stability (operating over 100 h), and superior rate properties ranging from 1 to 500 mA cm^{-2}.

More recently, Yang *et al.* reported a novel 3D nitrogenrich carbon architecture via a dual-templating strategy (Fig. 7).[124] By using both silica and ZIF-8 as templates, a 3D architecture of ordered macroporous interconnecting frameworks with thin mesoporous/microporous walls was obtained. This 3D architecture has an extraordinarily large surface area of 2546 m^2 g^{-1} and a high nitrogen doping content of 7.6 at%. The obtained Zn–air batteries displayed a high capacity of 770 mAh g^{-1} at a current density of 120 mAc m^{-2}. Furthermore, it exhibited a high power density of 197 mW cm^{-2} (three times of the value for the device based on 20 wt% Pt/C) at low mass loading of 0.5 mg cm^{-2}. The superior performance of the 3D carbon structure architectures may be attributed to their unique pore structures, ultrahigh surface area along with *in situ* nitrogen doping.

3. Conclusion and Prospects

Developing advanced electrode materials is of great importance for next-generation rechargeable batteries. Recently, MOFs have been utilized as precursors or sacrificial templates to prepare various nanostructured materials, such as porous carbon, metal oxides, other metal compounds and their composites. MOF-derived materials have large surface area, narrow distributed pore size, abundant porosity and designable structures, which made them potential candidates as the electrode of high-performance post-lithium-ion batteries. In this mini review, the preparation methods, various design strategies and the structure–performance relationships of recently reported MOF-derived materials have been summarized. Previous reports have shown that the morphology and composition of MOFs, pyrolysis conditions, post-treatment process and hybridization with other functional materials are beneficial to enhance their performance for rechargeable batteries applications. For example, porous carbon structures derived from MOFs with properly designed pore structure and composition have shown promising performance as cathode of Li–S batteries. The ultrahigh porosity of MOF-derived carbon enables a high sulfur loading. Meanwhile, narrow distributed micropores can effectively immobilize sulfur, leading to excellent cycling stability. Constructing hierarchically porous structures can effectively enhance ion diffusion and thus improve rate performance of fabricated electrodes. Despite the rapid development of MOF-derived materials, more efforts need be devoted to realize their practical applications. Currently, most of the works focus on using some common types of MOFs, such as ZIFs and MILs as precursors. Exploiting new MOFs by designing at the molecular scale is highly desirable in order to achieve precisely controlled structures and properties. Besides, constructing more complex architectures at the macroscopic level is of great potential. Particularly, using MOFs as building blocks to create high-order superstructures may open up a new avenue to improve the performance of their derived materials. Overall, the recent researches of MOF-derived materials as electrodes of next-generation rechargeable batteries are summarized. We hope this mini review can provide some new strategies for exploring MOF-derived materials in future.

Acknowledgments

This work was supported by the financial support from the Natural Science Foundation of Zhejiang Province (LQ17B030002, LR19E020013) and the National Natural Science Foundation of China (51702286). X. Cao thanks the financial support from the National Natural Science Foundation of China (51602284) and the

"Thousand Talent Program" and "Qianjiang Scholars" program of Zhejiang Province in China. The first three authors contribute equally to this work.

References

1. S. Zheng *et al.*, *Adv. Energy Mater.* **7**, 1602733 (2017).
2. Y. Jiang *et al.*, *Small* **14**, 1704296 (2018).
3. J. Ni and L. Li, *Adv. Funct. Mater.* **28**, 1704880 (2018).
4. Y. Zhong *et al.*, *Adv. Funct. Mater.* **28**, 1706391 (2018).
5. X. Cao *et al.*, *Adv. Mater.* **27**, 4695 (2015).
6. M. Huang *et al.*, *J. Mater. Chem. A* **5**, 266 (2017).
7. F. Zheng *et al.*, *Nanoscale* **7**, 9637 (2015).
8. G. Zhang *et al.*, *Adv. Mater.* **27**, 2400 (2015).
9. H. Wang *et al.*, *J. Mater. Chem. A* **5**, 23221 (2017).
10. R. Wu *et al.*, *Adv. Mater.* **27**, 3038 (2015).
11. W. Shi *et al.*, *Sci. Rep.* **6**, 18966 (2016).
12. Y. Zhong *et al.*, *Adv. Energy Mater.* **8**, 1701110 (2018).
13. H. Jiang *et al.*, *Angew. Chem. Int. Ed.* **57**, 3916 (2018).
14. J. Ni *et al.*, *Adv. Mater.* **29**, 1605607 (2017).
15. W. Li *et al.*, *Adv. Mater.* **29**, 1605820 (2017).
16. X. Cao *et al.*, *Adv. Mater.* **28**, 6167 (2016).
17. M. Zhang *et al.*, *Adv. Mater.* **30**, 1705431 (2018).
18. I. S. Amiinu *et al.*, *Adv. Funct. Mater.* **27**, 1702300 (2017).
19. Z. Luo *et al.*, *Small* **12**, 5920 (2016).
20. J. Ping *et al.*, *Adv. Mater.* **28**, 7640 (2016).
21. L. Zhang *et al.*, *Chem. Eur. J.* **24**, 13792 (2018).
22. X. Cao *et al.*, *Angew. Chem. Int. Ed.* **53**, 1404 (2014).
23. R. Demir-Cakan *et al.*, *J. Am. Chem. Soc.* **133**, 16154 (2011).
24. X. Xu *et al.*, *Adv. Funct. Mater.* **28**, 1707573 (2018).
25. M. Yang *et al.*, *Adv. Funct. Mater.* **27**, 1701971 (2017).
26. P. Li *et al.*, *Nanotechnology* **29**, 445401 (2018).
27. W. Liu *et al.*, *Angew. Chem. Int. Ed.* **56**, 5512 (2017).
28. X. Cao *et al.*, *Chem. Soc. Rev.* **46**, 2660 (2017).
29. Y. Wang *et al.*, *Adv. Mater.* **28**, 4149 (2016).
30. Z. Chen *et al.*, *J. Catal.* **361**, 322 (2018).
31. X. Shi *et al.*, *Mater. Lett.* **161**, 332 (2015).
32. G. Zou *et al.*, *Electrochim. Acta* **196**, 413 (2016).

33. Z. Chen *et al.*, *ACS Appl. Mater. Interfaces* **10**, 7134 (2018).
34. R. Wu *et al.*, *J. Power Sources* **330**, 132 (2016).
35. X. Xu *et al.*, *Chem. Mater.* **29**, 6058 (2017).
36. D. Sun *et al.*, *ACS Appl. Mater. Interfaces* **9**, 5254 (2017).
37. J. Li *et al.*, *J. Mater. Chem. A* **5**, 20428 (2017).
38. Y. Zhang *et al.*, *ACS Appl. Mater. Interfaces* **9**, 3624 (2017).
39. Z. Li *et al.*, *Nano Energy* **32**, 494 (2017).
40. J. Qian *et al.*, *Chem. Commun.* **53**, 13027 (2017).
41. Q. Lai *et al.*, *Small* **13**, 1700740 (2017).
42. F. Cao *et al.*, *J. Am. Chem. Soc.* **138**, 6924 (2016).
43. X. Bai *et al.*, *Chemistry* **23**, 14839 (2017).
44. G. Yilmaz *et al.*, *Adv. Mater.* **29**, 1606814 (2017).
45. W. Yang *et al.*, *Chem. Commun.* **53**, 12934 (2017).
46. D. Xiao *et al.*, *J. Mater. Chem. A* **5**, 24901 (2017).
47. Z. Li *et al.*, *Nano Energy* **23**, 15 (2016).
48. M. Zhang *et al.*, *Adv. Mater.* **30**, 1705431 (2018).
49. R. Li *et al.*, *J. Hazard Mater.* **338**, 167 (2017).
50. Y. Tong *et al.*, *Angew. Chem. Int. Ed.* **56**, 7121 (2017).
51. C.-Y. Su *et al.*, *Adv. Energy Mater.* **7**, 1602420 (2017).
52. H. Zhang *et al.*, *J. Mater. Chem. A* **6**, 7133 (2018).
53. A. Y. Kim *et al.*, *ACS Appl. Mater. Interfaces* **8**, 19514 (2016).
54. X. Zhang *et al.*, *Chem. Commun.* **51**, 16413 (2015).
55. F. Zou *et al.*, *ACS Nano* **10**, 377 (2016).
56. C. Wang *et al.*, *Mater. Horiz.* **5**, 394 (2018).
57. H. Pang *et al.*, *Electrochim. Acta* **213**, 351 (2016).
58. X. Yang *et al.*, *J. Mater. Chem. A* **3**, 15314 (2015).
59. Z. Chen *et al.*, *J. Mater. Chem. A* **6**, 10304 (2018).
60. Z. Chen *et al.*, *Chem. Eng. J.* **326**, 680 (2017).
61. G. Fang *et al.*, *J. Mater. Chem. A* **5**, 13983 (2017).
62. W. Xia *et al.*, *Chem. Commun.* **54**, 1623 (2018).
63. C. Guan *et al.*, *Nanoscale Horiz.* **2**, 99 (2017).
64. C. Young *et al.*, *Chem. Mater.* **30**, 3379 (2018).
65. X. Hu *et al.*, *ACS Appl. Mater. Interfaces* **10**, 14684 (2018).
66. X. Xu *et al.*, *J. Mater. Chem. A* **4**, 6042 (2016).
67. Z. Chen *et al.*, *Nano Res.* **11**, 966 (2017).
68. J. Xu *et al.*, *J. Mater. Chem. A* **6**, 2797 (2018).
69. D. Li *et al.*, *J. Colloid. Interface Sci.* **497**, 350 (2017).
70. D. Ji *et al.*, *Chem. Eng. J.* **313**, 1623 (2017).
71. Z. W. Seh *et al.*, *Chem. Soc. Rev.* **45**, 5605 (2016).

72. X. J. Hong *et al.*, *Nanoscale* **10**, 2774 (2018).
73. X. J. Hong *et al.*, *ACS Appl. Mater. Interfaces* **10**, 9435 (2018).
74. H. B. Wu *et al.*, *Chem. Eur. J.* **19**, 10804 (2013).
75. S. Bai *et al.*, *Nat. Energy* **1**, 16094 (2016).
76. G. Xu *et al.*, *J. Mater. Chem. A* **1**, 4490 (2013).
77. Z. Li and L. Yin, *ACS Appl. Mater. Interfaces* **7**, 4029 (2015).
78. Y. Jiang *et al.*, *ACS Appl. Mater. Interfaces* **9**, 25239 (2017).
79. X. Qian *et al.*, *RSC Adv.* **6**, 94629 (2016).
80. L. Wang *et al.*, *Coordin. Chem. Rev.* **307**, 361 (2016).
81. J. He *et al.*, *J. Mater. Chem. A* **6**, 10466 (2018).
82. Y. Guo *et al.*, *Mater. Lett.* **199**, 101 (2017).
83. C. Li *et al.*, *Sci. Rep.* **6**, 25556 (2016).
84. X. Shi *et al.*, *J. Power Sources* **330**, 1 (2016).
85. W. Tang *et al.*, *J. Mater. Chem. A* **4**, 4882 (2016).
86. X. Song *et al.*, *Nano Energy* **37**, 90 (2017).
87. G. Ali *et al.*, *ACS Appl. Mater. Interfaces* **8**, 15422 (2016).
88. J. Xu *et al.*, *Nano Energy* **50**, 323 (2018).
89. Q. Zheng *et al.*, *J. Mater. Chem. A* **6**, 4209 (2018).
90. F. Li *et al.*, *J. Mater. Chem. A* **5**, 25276 (2017).
91. R. J. Clément *et al.*, *Chem. Mater.* **28**, 8228 (2016).
92. J. Billaud *et al.*, *J. Am. Chem. Soc.* **136**, 17243 (2014).
93. L.-W. Jiang *et al.*, *Chin. Phys. Lett.* **35**, 048801 (2018).
94. Z. Zheng *et al.*, *J. Mater. Sci.* **53**, 12421 (2018).
95. P. Liu *et al.*, *J. Power Sources* **395**, 158 (2018).
96. Y. Li *et al.*, *Energy Environ. Sci.* **9**, 1998 (2016).
97. D. Su *et al.*, *Adv. Mater.* **29** 1700587 (2017).
98. J. Zhou *et al.*, *Energy Environ. Sci.* **7**, 2715 (2014).
99. Y. Mao *et al.*, *Nat Commun.* **8**, 14628 (2017).
100. X. Lu *et al.*, *Mater. Res. Bull.* **99**, 45 (2018).
101. L. Zhou *et al.*, *Nano Energy* **35**, 281 (2017).
102. G. Fang *et al.*, *J. Mater. Chem. A* **5**, 13983 (2017).
103. K. Zhang *et al.*, *Adv. Funct. Mater.* **26**, 6728 (2016).
104. S. Dong *et al.*, *ACS Nano* **11**, 6474 (2017).
105. Y. Wang *et al.*, *J. Mater. Chem. A* **4**, 5428 (2016).
106. X. Ge *et al.*, *Nano Energy* **32**, 117 (2017).
107. X. Zhang *et al.*, *Ceram. Int.* **43**, 2398 (2017).
108. G. Zou *et al.*, *J. Power Sources* **325**, 25 (2016).
109. X. Zhang *et al.*, *J. Colloid Interface Sci.* **499**, 145 (2017).
110. L. Yu *et al.*, *ACS Nano* **11**, 5120 (2017).

111. G. Zou *et al.*, *Small* **14**, 1702648 (2018).

112. X. Zhang *et al.*, *J. Alloy. Compd.* **710**, 575 (2017).

113. S. Wang *et al.*, *Nano Energy* **39**, 626 (2017).

114. Y. Qian *et al.*, *Carbon* **111**, 641 (2017).

115. L. Yang *et al.*, *Nano Energy* **50**, 691 (2018).

116. H. Han *et al.*, *ChemElectroChem* **5**, 1868 (2018).

117. T. Wang *et al.*, *Adv. Mater.* **30**, 1800757 (2018).

118. Y. Jiang *et al.*, *Adv. Energy Mater.* **8**, 1702900 (2018).

119. J. Zhu *et al.*, *Small* **14**, 1800563 (2018).

120. Q. Wang *et al.*, *ACS Energy Lett.* **3**, 1183 (2018).

121. Y. Qian *et al.*, *Small* **13**, 1701143 (2017).

122. J. Liu *et al.*, *Adv. Energy Mater.* **7**, 1700518 (2017).

123. C. Lin *et al.*, *J. Mater. Chem. A* **5**, 13994 (2017).

124. M. Yang *et al.*, *Adv. Funct. Mater.* **27**, 1701971 (2017).

Chapter 14

Polymer-in-Salt Solid Electrolytes for Lithium-Ion Batteries

Chengjun Yi*, Wenyi Liu*, Linpo Li[†], Haoyang Dong*, and Jinping Liu*,[‡]

Solid-state polymer lithium-ion batteries with better safety and higher energy density are one of the most promising batteries, which are expected to power future electric vehicles and smart grids. However, the low ionic conductivity at room temperature of solid polymer electrolytes (SPEs) decelerates the entry of such batteries into the market. Creating polymer-in-salt solid electrolytes (PISSEs) where the lithium salt contents exceed 50 wt.% is a viable technology to enhance ionic conductivity at room temperature of SPEs, which is also suitable for scalable production. In this review, we first clarify the structure and ionic conductivity mechanism of PISSEs by analyzing the interactions between lithium salt and polymer matrix. Then, the recent advances on polyacrylonitrile (PAN)-based PISSEs and polycarbonate derivative-based PISSEs will be reviewed. Finally, we propose possible directions and opportunities to accelerate the commercializing of PISSEs for solid polymer Li-ion batteries.

*School of Chemistry, Chemical Engineering and Life Science, State Key Laboratory of Advanced Technology for Materials Synthesis and Processing, Wuhan University of Technology, Wuhan 430070, P. R. China.
[†]School of Optical and Electronic Information, Huazhong University of Science and Technology, Wuhan 430074, P. R. China.
[‡]liujp@whut.edu.cn

Keywords: Solid polymer electrolytes; polymer in salt; solid-state batteries; mechanical properties.

1. Introduction

Lithium-ion batteries (LIBs) have been extensively studied and widely applied by virtue of their long cycle life, high energy density and minimal environmental pollution, since they were commercialized by the Sony Corporation in the 1990s.[1-3] Unfortunately, state-of-the-art LIBs are still trapped in safety concerns and the bottleneck of energy density due to the use of liquid electrolytes that are susceptible to leakage and flammability. Thus, to fabricate next-generation LIBs featured with higher energy density and safety for electric vehicles and smart grids, replacing liquid electrolytes with solid polymer electrolytes (SPEs) is a promising route.[4-6] In general, SPEs exhibit distinct advantages such as nonflammability, flexibility, a wide electrochemical window and so on. However, their ionic conductivities at room temperature are too low to be practical.[7] Among various methods to improve ionic conductivities at room temperature, forming polymer-in-salt solid electrolytes (PISSEs) by adding high-concentration lithium salt is a facile and large-scale production strategy.[8,9] Therefore, it is highly necessary to investigate the PISSEs for developing next-generation LIBs with higher energy density and safety.

First of all, to clarify the design purpose of PISSEs, conventional SPEs are introduced. In 1973, Wright's[10] group discovered that polyethylene oxide (PEO) complexed with alkali metal salts had the property of ionic conductivity. PEO could dissolve lithium salt by coordination between ether group in the polymer chains and Li^+, and then transfer Li^+ by the local relaxation and segmental motion of polymer chains under an electric field.[11] Hence, Li^+ density in the electrolyte and the glass transition temperature (T_g) of the polymer are crucial for the ionic conductivity of SPEs. In 1992, Angell et al.[12] found that T_g would reach a maximum with the higher salt concentrations, where increasing the salt concentrations would increase not only the ionic mobility but also the freedom of Li^+ to move independently of their surroundings. In 1993, Angell and coworkers introduced an innovative polymer system described as "polymer in salt" by reversing the ratio of the polymer matrix to lithium salt, which delivers good ionic conductivity and high electro-chemical stability.[8]

Although PISSEs have high ionic conductivities at room temperature and good electrochemical stability, their mechanical properties are relatively poor. It is undeniable that the mechanical performances are deteriorated due to ultrahigh concentration of lithium salt in PISSEs. To address this issue, some attempts have been devoted to selecting polymer matrix with good mechanical properties or adding materials with high mechanical strength into the PISSEs.[13,14]

Herein, the review is focused on surveying the current development and issues of PISSEs for LIBs. First, we specially review the possible structure and ion-conduction mechanism of PISSEs; then we will highlight recent progress on PISSEs; finally, we will end this review with a discussion about the current limitations, remaining challenges and prospects for future research of PISSEs.

2. Structure and Ion-Conduction Mechanisms of PISSEs

To expose the structure details and further interpret the ion-conduction mechanism of PISSEs, a key is to understand the interactions between the polymer matrix and the lithium salt. Before forming PISSEs, SPEs have experienced three structural changes due to different interactions between the polymer and the lithium salt in various amounts:[15] (1) at low salt content, only Li^+ of the dissolved salt associates with the polymer; (2) ionic associates begin to appear and some of them form aggregates with the polymer when adding to more salts; (3) a few ionic clusters and transient polymer segments composed of ionic clusters with the polymer host are created when the salt contents are high. In general, the PISSEs will be formed when the content of lithium salt exceeds 50% in SPEs. Under such a condition, the structure of PISSEs mainly contains an efficient percolation path constructed by infinitely connected ionic clusters as well as the complexes.[16,17]

Ion-conduction mechanisms of PISSEs are closely related to the above structures. For common SPEs, the ionic migration has to only depend on the segmental motion of the polymer matrix, causing low ionic conductivities due to high T_g of the polymer. For PISSEs, additionally, the species of the ionic groups complexed with the polymer are also important for the ionic conductivities. Li^+, ionic associates and ionic clusters are all able to move on the polymer chain by repeatedly coordinating/dissociating with the polar groups in the polymer matrix. In PISSEs, besides, the ionic transference via hopping dominantly occurs in the limitless network of connected ionic clusters. Hence, the ionic conductivity of PISSEs is several orders of magnitude higher than that of common SPEs.[15]

3. Recent Advances on PISSEs

To form high-concentration salt PISSEs, the dissociation energy of the lithium salt must be low. Lithium bis (trifluoromethane sulfone) imide (LiTFSI) and lithium bis (fluorosulfonyl)imide (LiFSI) are often used as lithium salts for PISSEs system, because they are easily dissolved in the polymer owing to high delocalization of charge in the anions.[18] As another essential component, the polymer matrix not only enables dissolution of a considerable amount of lithium salt, but also acts as the framework of the electrolyte membrane. Thus, the polymer backbone should have a high dielectric constant and good mechanical strength. According to the polymer host, current PISSEs can be mainly divided into polyacrylonitrile (PAN)-based PISSEs and polycarbonate derivative-based PISSEs.

3.1. *PAN-based PISSEs*

Among PAN-based PISSEs systems, the PAN-LiTFSI PISSE is the most popular.[15,16,19] However, the system has low ionic conductivity and poor electrochemical stability.[20] This is because of vigorous CN groups in PAN, causing high thermodynamic but low kinetic flexibility of the chains.[21] In order to further enhance ionic conductivity and improve electrochemical stability, adding inorganic functional materials to PISSEs is useful. Some inorganic materials, with intrinsic stability, can inhibit polymer crystallization to gain more flexible chains and optimize the interaction between polymer and Li⁺.[22] They can also improve other properties of PISSEs, like mechanical strength and thermal stability, etc.

Wu *et al.*[23] reported a PAN-LiTFSI SPE (84 wt.% LiTFSI in 16 wt.% PAN) incorporated with graphene oxide (GO) nanosheets (Fig. 1(a)), which improved the ionic conductivity, electrochemical stability, thermal stability, and chemical stability of the SPE. The SPE with 0.9 wt.% GO nanosheets shows the ionic conductivity of $1:1 \times 10^{-4}$ S cm⁻¹ at 30°C, which is almost one order of magnitude higher than that of the filler-free one (Fig. 1(b)). This is because these

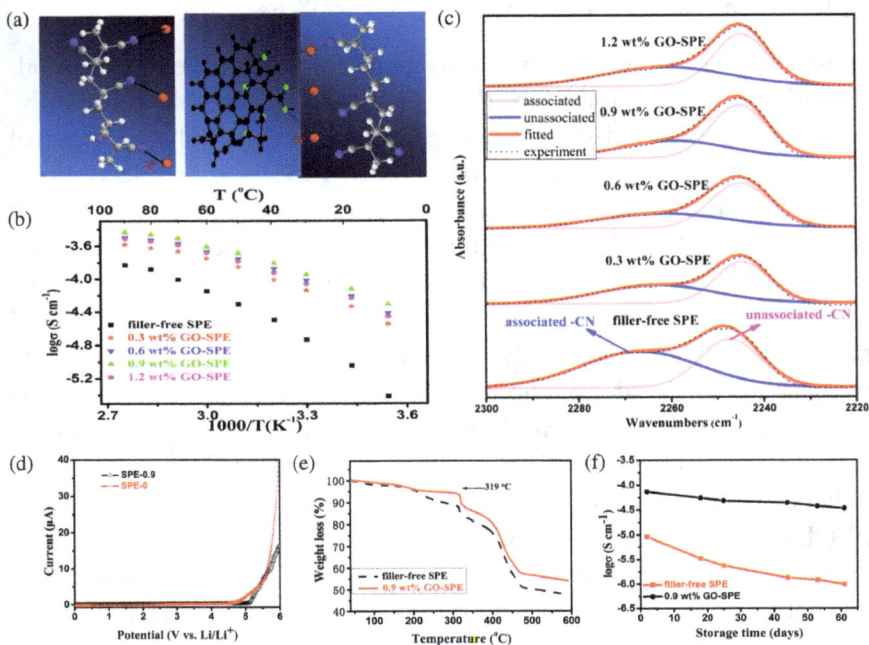

Fig. 1. (a) Schematic illustration of GO effect on SPE. (b) Arrhenius plots of the composite polymer electrolytes with various GO loadings. (c) FTIR spectra of SPE with various GO contents. (d) LSV of Li/SPE-*x*/SS cells at 1 mVs⁻¹. (e) TGA curves of pristine SPE and 0.9 wt.% GO-SPE. (f) Bulk conductivity of Li/SPE-*x*/Li cells with storage time at room temperature. Reproduced with permission from Ref. 23. Copyright 2016: The Electrochemical Society.

oxygencontaining groups (e.g., –COOH, –CO, –OH and –COC–) in laminated GO interact with Li$^+$, leading to the decoupling of the ionic transport from segmental motion in the electrolyte and building up a conductive framework on the surface of GO nanosheets (Fig. 1(c)). Moreover, the electrochemical window is expanded by 0.5 V to approximately 5 V (vs. Li$^+$/Li); and the thermal stability is also better than pristine SPE attributed to an isolating effect between PAN and GO nanosheets (Figs. 1(d) and 1(e)). Wu *et al.* also found that GO nanosheets could maintain the bulk conductivity of the PAN-LiTFSI SPE by avoiding the metaphase possibly formed via high-concentration LiTFSI crystallization and precipitation from PAN (Fig. 1(f)).

3.2. *Polycarbonate derivative-based PISSEs*

In recent years, polycarbonates have been found to have high solubility and good dissociation for lithium salts and are suitable as the polymer matrix for solid-state LIBs, especially as the polymer host of PISSEs.[24–26] However, polycarbonatebased PISSEs are subjected to a serious deterioration of mechanical properties, which probably causes failure to form a self-standing film. To boost the mechanical properties, adding materials with excellent mechanical strength is a simple and efficient method. Tominaga's[27] group first synthesized poly(ethylene carbonate) (PEC)-based PISSE with 80 wt.% LiFSI by using a 3D ordered macroporous polyimide matrix to improve its mechanical properties. However, the hybrid membrane sacrifices parts of the ionic conductivity of PEC. Whereafter, Zhao *et al.*[28] put poly (vinylidene fluoride-co-hexafluoropropylene) (PVDF-HFP) into PECLi80 electrolyte (80 wt.% LiTFSI in 20 wt.% PEC) with polymer-in-salt structure to preserve the conductivity of PEC and enhance the mechanical performance of the polymer electrolyte using the better mechanical strength of PVDF-HFP (Fig. 2). The new hybrid polymer electrolyte FPEC80/50 (50 wt.% PEC80 in 50 wt.% PVDF-HFP) exhibits not only a high ionic conductivity of $1{:}08 \times 10^{-4}$ S cm^{-1} at 30°C, but also a stable electrochemical window around 4.5 V (vs. Li$^+$/Li) due to the synergistic effect between PEC and PVDF-HFP

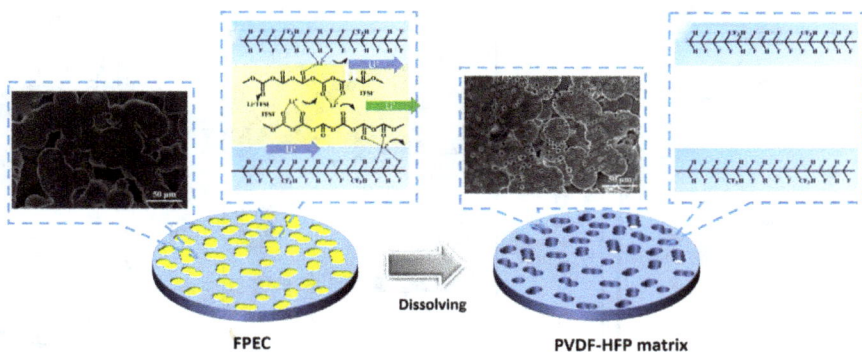

Fig. 2. Schematic illustration of FPEC. Reproduced with permission from Ref. 28. Copyright 2018: Elsevier.

(Figs. 3(a)–3(c)). As a result, the Li/FPEC80/50/LiFePO$_4$ cell displays excellent rate capacity and cycling stability (Figs. 3(d)–3(f)).

To manufacture higher energy density batteries, solid-state lithium metal batteries must be considered, such as Li-S[29,30] and Li-O^2 (Refs. 31 and 32) batteries. This is because lithium metal renders an ultrahigh specific capacity (3860 mAh g^{-1}) and the most negative electrochemical potential (−3.04 V vs. standard hydrogen electrode). Whereas, the growth of lithium dendrites will puncture the electrolyte membrane and cause a short circuit when cycling, especially at high current densities. Thus, if applied in such cases, the PISSE needs a quite high mechanical rigidity for inhibiting lithium dendrites.

Chen *et al.*[33] proposed a polymer-in-salt bi-grafted polysiloxane (BPSO) copolymer electrolyte (the mass ratio of LiTFSI to BPSO is 150%) with superhigh ionic conductivity at room temperature, and further prepared the composite polymer electrolyte membrane with high ionic conductivity and good mechanical strength by simultaneously selecting cellulose acetate (CA) membranes as the backbone and polymer-in-salt BPSO copolymers as the ion-conducting materials. After introducing CA, although the ionic conductivity of BPSO copolymer electrolyte is sacrificed, its mechanical strength is remarkably improved; thus the lithium dendrites growth is blocked (Figs. 4(a)–4(e)).

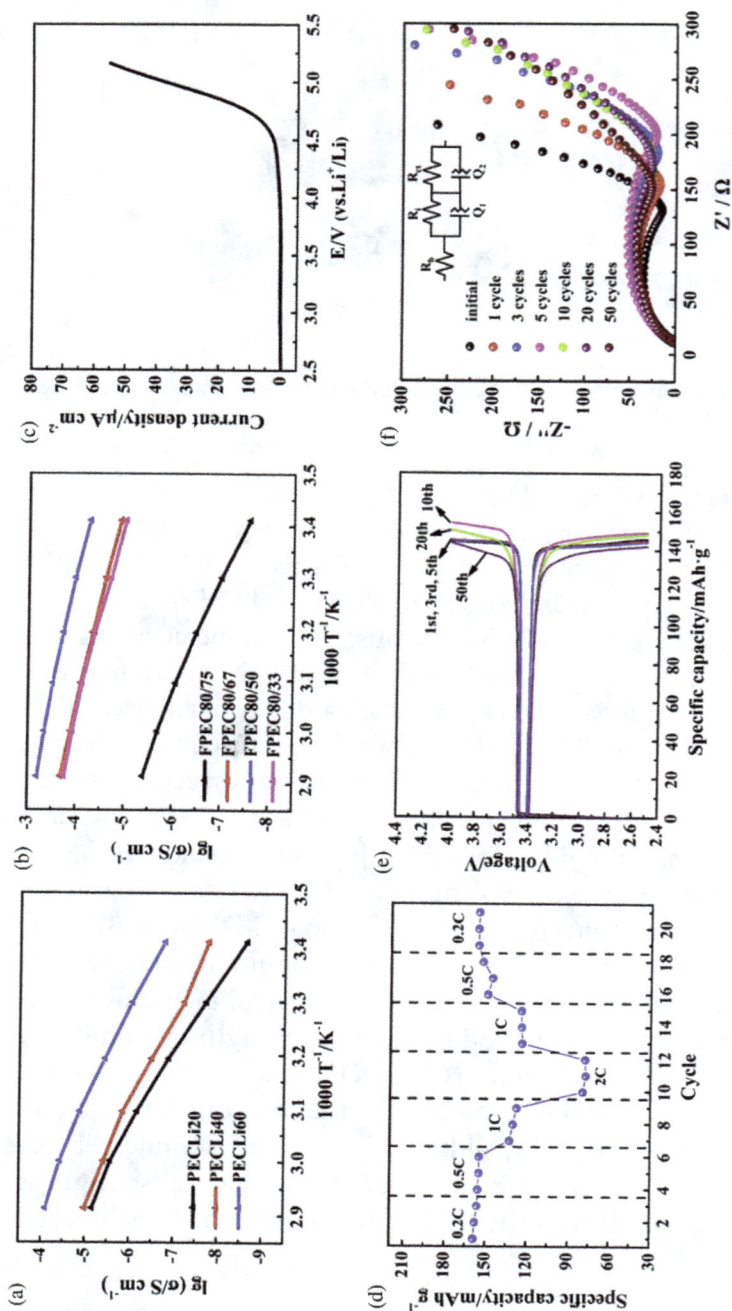

Fig. 3. The temperature-dependent conductivity of (a) PECLi with different Li salt concentrations and (b) FPEC80 with different ratios of PVDF-HFP. (c) Electrochemical stability window of FPEC80/50 which measured using LSV that scans at 1 mV s⁻¹. (d) Rate capability of LFP/FPEC/Li cell. (e) Charge/Discharge curves of LFP/FPEC80/50/Li cell in different cycles. (f) Variation in AC impedance spectra of LFP/FPEC/Li cell in different test cycle at 50°C (the inset is the fitting circuit for semicircle). Reproduced with permission from Ref. 28. Copyright 2018: Elsevier.

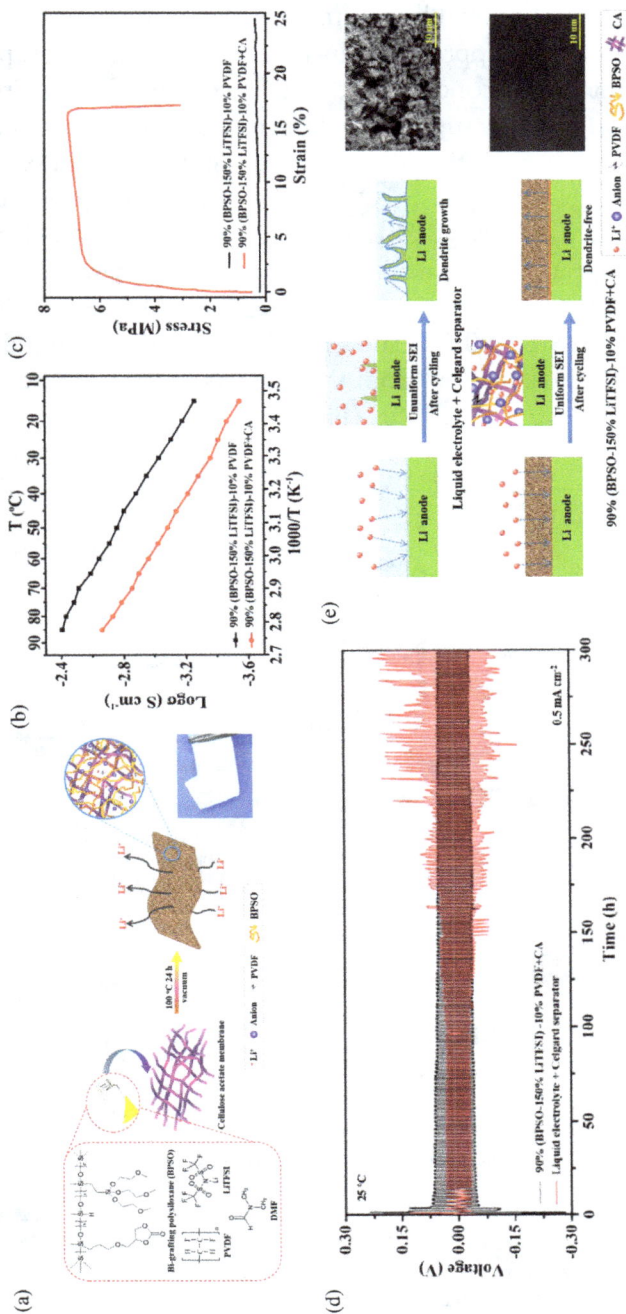

Fig. 4. (a) Schematic illustration of fabricating composite polymer electrolyte membrane by solution-casting technique. (b) Temperature dependences of ionic conductivities for 90% (BPSO-150% LiTFSI)-10% PVDF and 90% (BPSO-150% LiTFSI)-10% PVDF + CA membranes. (c) Stress–strain curves. (d) Comparison of constant current cycling curves of 90% (BPSO-150% LiTFSI)-10% PVDF + CA and liquid electrolyte + Celgard separator at 0.5 mA cm⁻² in lithium symmetric cell. (e) Schematic illustrations of Li plating/stripping behaviors lithium symmetric cells using liquid electrolyte + Celgard separator and 90% (BPSO-150% LiTFSI)-10% PVDF + CA. Reproduced with permission from Ref. 33. Copyright 2018: Elsevier.

Introducing active ceramic fillers into polycarbonate derivative-based PISSEs is a practicable approach to increase the mechanical properties as well as the ionic conductivity. Active ceramic fillers with inherent hardness can not only suppress the polymer crystallization, but also conduct Li^+.[34,35] Li *et al.*[36] fabricated a ceramic/polymer composite electrolyte by dissolving $Li_{1.5}Al_{0.5}Ge_{1.5}(PO_4)_3$ (LAGP) into poly(propylene carbonate) (PPC)-LiFSI 80 wt.% PISSE solution. This PPC-based electrolyte was considered as an adequate alternative of PEO-based composed polymer electrolyte. Compared with LAGP/PEO-SPE-80/20 electrolyte, LAGP/LiFSI80-80/20 shows 1–2 orders of magnitude higher ionic conductivity beneficial from the synergistic effect between LAGP and PPC-LiFSI 80 wt.% PISSE (Fig. 5(a)). Consequently, the

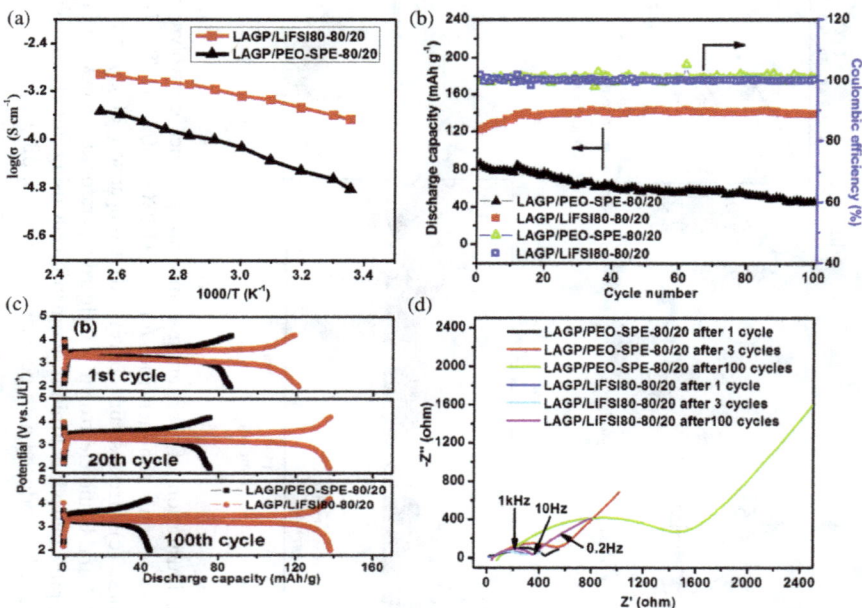

Fig. 5. (a) Arrhenius plots of ionic conductivity for electrolytes LAGP/PEO-SPE-80/20 and LAGP/LiFSI80-80/20. Cycling performance (b) and charge/discharge curves (c) of $LiFePO_4$/Li cells with electrolytes LAGP/PEO-SPE-80/20 and LAGP/LiFSI80-80/20 at 0.1 C, 25°C. (d) Electrochemical impedance spectra of 100% DOC $LiFePO_4$/Li cells with LAGP/PEO-SPE-80/20 and LAGP/LiFSI80-80/20 electrolytes at 25°C after 1, 3 and 100 cycles, respectively. Reproduced with permission from Ref. 36. Copyright 2018: Elsevier.

Li/LAGP/LiFSI80-80/20/LiFePO$_4$ cell displays better cycling property than the Li/LAGP/PEO-SPE-80/20/LiFePO$_4$ cell, which is also due to the improved interfacial behavior during cycling between electrode and composite electrolyte (Figs. 5(b)–5(d)).

4. Perspective

Here, we propose the possible directions and opportunities to accelerate the commercializing of PISSEs for solid polymer LIBs.

4.1. *New lithium salts and polymer matrixes*

The cost of the lithium salt must be taken into account due to the massive use in PISSEs. Whereas, lithium salts used for PISSEs systems are either LiFSI or LiTFSI, which are both expensive. Furthermore, the high concentration of lithium salt is unstable by cause of easy crystallization.[37] Thus, it is necessary to develop new lithium salts with low cost and high stability. Unfortunately, few relevant investigations have been reported. Moreover, the anion of the new type of lithium salts should be multivalent and large. This is because of the following reasons: First, lithium salts composed of the large anion and Li$^+$ are readily dissociated in the polymer backbone.[38,39]

Second, the large anion may plasticize polymer chains to facilitate segment movement, which is beneficial for Li$^+$ transference. Third, the polyvalent anion can provide more Li$^+$ in order to maintain charge conservation, thus increasing Li$^+$ density and enhancing transference number of Li$^+$.

To further reduce the cost of PISSEs, it is of necessity to find some new polymer matrix. In addition to being cheap, the polymer host should also have the following characteristics: (1) high dielectric constant to boost the dissolution of lithium salts; (2) superior mechanical strength; (3) excellent stability, including thermal stability, chemical stability and electrochemical stability; (4) particular group that can immobilize the anion of the lithium salt to form single-ion conductivity, as Rolland *et al.* reported[40] (Fig. 6). Given the abundance and the mature industries, natural polymers and modified natural polymers are potential candidates.

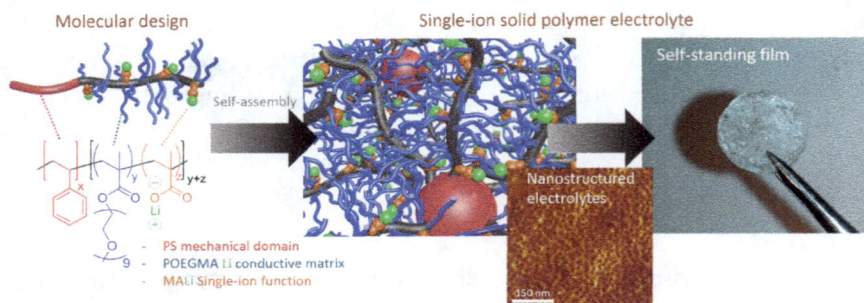

Fig. 6. Schematic illustration of single-ion SPE. Reproduced with permission from Ref. 40. Copyright 2018: Elsevier.

4.2. *Electrode-supported PISSE integrated membrane*

Although the properties of PISSEs can be improved by adding functional inorganic or organic materials, the performances of PISSEs-based LIBs still need to be improved. This is because of the poor interfacial contact between the polymer electrolyte and the electrode, which leads to huge interfacial resistance.[41-43] It is difficult for the rigid solid electrolyte membrane to effectively wet interface like liquid electrolytes. This issue can be addressed by casting the polymer electrolyte solution on the electrode to *in situ* form the polymer electrolyte membrane, constructing an electrode-supported PISSE integrated membrane[44-46] (Fig. 7). This design of the combined structure can not only unlock the performance of the electrode by increasing the contact area between the electrode and the solid electrolyte, but also enhance the energy density of the battery by reducing the thickness of the polymer electrolyte. Moreover, such solidstate battery can power flexible electronics, because it does not cause an increase in interfacial impedance between the electrode and the polymer electrolyte even under bending or deforming. To further promote the performances of the flexible battery with the electrode-supported PISSE integrated membrane, 3D electrode structure will be properly useful (Fig. 8), especially 3D nanoarray structure grown directly on a current collector.[47,48] The electrode with 3D nanoarray structure can achieve not only the rapid transport of electrons, but also the full

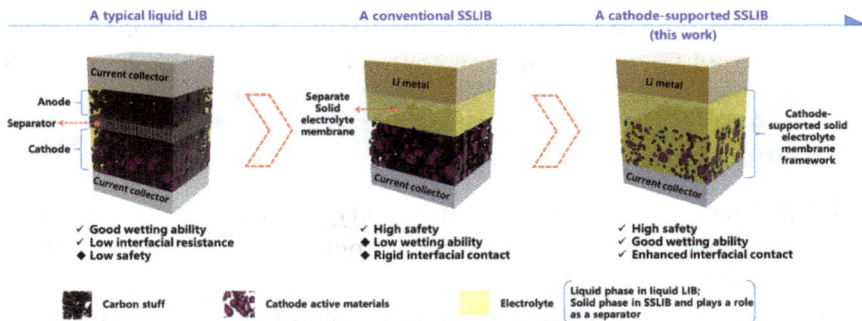

Fig. 7. Schematic of the novel cathode-supported SSLIB in comparison with a conventional rigid SSLIB and a typical liquid LIB. Reproduced with permission from Ref. 45. Copyright 2018: The Royal of Society of Chemistry.

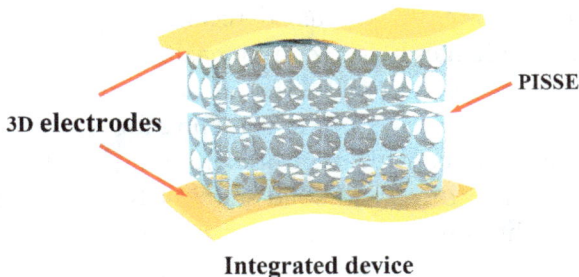

Fig. 8. Schematic illustration of a conceptual flexible device using 3D electrodes integrated with PISSE.

contact between the active materials and PISSEs thanks to the enough interspacing between the individual nanostructures. We further propose two strategies to realize the 3D integrated electrode/electrolyte structure, that is, physical and chemical methods. Physical method is facile, using which the polymer electrolyte solution is directly filled into 3D electrode to form electrolyte membrane, like casting,[45] spraying and so on. This method is suitable for realizing most integrated structures, but the degree of integration cannot be easily controlled. Chemical method specifically refers to *in situ* electrically[49] or chemically polymerizing of a monomer, or *in situ* cross-linking of a polymer within the 3D electrode. In contrast, chemical method can easily control the degree of integration as long as the polymer monomers or polymers used for polymerization/cross-linking are available.

5. Conclusion

Solid polymer LIBs with reassuring safety, high energy density and arbitrary shape show great potential for next-generation energy storage. To achieve commercially available solid polymer LIBs, SPEs with excellent performances, such as high ionic conductivity at room temperature, wide electrochemical window and sufficient mechanical properties are essential. Among various SPEs, PISSEs are promising due to superior ionic conductivity at room temperature. In this review, we have analyzed the microstructure and the ionconduction mechanism of PISSEs and discussed recent progress on improving PISSEs. Some challenges before commercialization like stability, cost, interfacial resistance between PISSEs and electrode, have also been put forward. In the coming studies, searching for optimal PISSEs and optimizing the interface with electrode are indispensable for advancing the performance of related solid-state LIBs. Developing PISSEs not only paves the way towards the commercialization of solid-state LIBs, but also sheds new light on the development of other advanced energy systems that employ polymer electrolytes. It is believed that great progress will be made soon in the near future.

Acknowledgments

This project was supported by the National Key R&D Program of China (2016YFA0202602) and the National Natural Science Foundation of China (51672205, 51872104).

References

1. J. M. Tarascon and M. Armand, *Nature* **414**, 359 (2001).
2. M. S. Whittingham, *Chem. Rev.* **104**, 4271 (2004).
3. Z. Xing *et al.*, *Funct. Mater. Lett.* **11**, 1850067 (2018).
4. R. J. Chen *et al.*, *Mater. Horiz.* **3**, 487 (2016).

5. Y. S. Hu, *Nat. Energy* **1**, 16402 (2016).
6. Z. G. Xue *et al.*, *J. Mater. Chem. A* **3**, 19218 (2015).
7. E. Quartarone and P. Mustarelli, *Chem. Rev.* **40**, 2525 (2011).
8. C. A. Angell *et al.*, *Nature* **362**, 137 (1993).
9. J. Fan and C. A. Angell, *Electrochim. Acta* **40**, 2397 (1995).
10. D. E. Fenton *et al.*, *Polymer* **14**, 589 (1973).
11. A. Magistris and K. Singh, *Polym. Int.* **28**, 277 (1992).
12. M. C. Mclin and C. A. Angell, *Solid State Ion.* **53/56**, 1027 (1992).
13. R. Khurana *et al.*, *J. Am. Chem. Soc.* **136**, 7395 (2014).
14. X. Zhang *et al.*, *J. Am. Chem. Soc.* **139**, 13779 (2017).
15. Z. X. Wang *et al.*, *Electrochem. Solid-State Lett.* **4**, A148 (2001).
16. Z. X. Wang *et al.*, *J. Electrochem. Soc.* **149**, E148 (2002).
17. H. X. Wang *et al.*, *Chem. Commun.* **10**, 2186 (2004).
18. M. Armand *et al.*, *Second Int. Symp. Polym. Electrolytes* (1990), pp. 91–97.
19. M. Forsyth *et al.*, *Electrochem. Acta* **45**, 1249 (2000).
20. F. Croce and B. Scrosati, *J. Power Sources* **43**, 9 (1993).
21. O. V. Bushkova *et al.*, *Solid State Ion.* **119**, 217 (1999).
22. C. Tang *et al.*, *Nano Lett.* **12**, 1152 (2012).
23. B. Wu *et al.*, *J. Electrochem. Soc.* **163**, A2248 (2016).
24. Y. Cui *et al.*, *Appl. Mater. Interfaces* **9**, 8737 (2017).
25. J. Zhang *et al.*, *J. Mater. Chem. A* **5**, 4940 (2017).
26. J. Zhang *et al.*, *Adv. Energy Mater.* **5**, 1501082 (2015).
27. K. Kimura *et al.*, *Electrochem. Commun.* **66**, 46 (2016).
28. Y. Zhao *et al.*, *J. Power Sources* **407**, 23 (2018).
29. S. Zhu and Y. Li, *Funct. Mater. Lett.* **11**, 1830007 (2018).
30. L. P. Chen *et al.*, *Funct. Mater. Lett.* **11**, 1840010 (2018).
31. L. Xiao *et al.*, *Sci. China Mater.* **62**, 633 (2019).
32. L. Xiao *et al.*, *Rare Metals* **37**, 527 (2018).
33. L. Chen and L. Z. Fan, *Energy Storage Mater.* **15**, 37 (2018).
34. W. Zhou *et al.*, *J. Am. Chem. Soc.* **138**, 9385 (2016).
35. R. Chen *et al.*, *ACS Appl. Mater. Interfaces* **9**, 9654 (2017).
36. Y. Li *et al.*, *J. Power Sources* **397**, 95 (2018).
37. A. K. Łasińska *et al.*, *Electrochim. Acta* **169**, 61 (2015).
38. L. Chen *et al.*, *Adv. Funct. Mater.* **29**, 1901047 (2019).
39. L. Jennifer *et al.*, *Chem. Mater.* **25**, 834 (2013).
40. J. Rolland *et al.*, *Polymer* **68**, 344 (2015).
41. V. Augustyn *et al.*, *Joule* **2**, 2189 (2018).
42. X. W. Yu and A. Manthiram, *Energy Environ. Sci.* **11**, 527 (2018).

43. J. M. Giussi *et al.*, *Chem. Soc. Rev.* **48**, 814 (2019).
44. W. H. Zuo *et al.*, *Adv. Sci.* **4**, 1 (2017).
45. X. Z. Chen *et al.*, *Energy Environ. Sci.* **12**, 936 (2018).
46. Q. Y. Xia *et al.*, *Small* **14**, 1804149 (2018).
47. J. Jiang *et al.*, *Nanoscale* **3**, 45 (2011).
48. J. Jiang *et al.*, *Adv. Mater.* **24**, 5166 (2012).
49. G. Salian *et al.*, *J. Power Sources* **340**, 242 (2017).